I0493816

Disclaimer

The publisher of this book is by no way associated with the National Institute of Standards and Technology (NIST). The NIST did not publish this book. It was published by 50 page publications under the public domain license.

50 Page Publications.

Book Title: Effect of Al2O3 Nanolubricant on R134a Pool Boiling Heat Transfer with Extensive Measurement and Analysis Details

Book Author: Mark A. Kedzierski;

Book Abstract: This paper quantifies the influence of Al2O3 nanoparticles on the pool boiling performance of R134a/polyolester mixtures on a roughened, horizontal, flat surface. Nanofluids are liquids that contain dispersed nano-size particles. A lubricant based nanofluid (nanolubricant) was made by suspending 20 nm diameter Al2O3 particles in a synthetic ester to roughly a 1.6 % volume fraction. The nanoparticles enhanced the boiling heat transfer relative to that for R134a/polyolester mixtures without nanoparticles for the three lubricant mass fractions that were tested. The enhancement occurred for the lowest heat fluxes, which gives the opportunity for designing chillers for lower approach temperatures. For the 0.5 % nanolubricant mass fraction, the nanoparticles caused a heat transfer enhancement relative to the heat transfer of pure R134a/polyolester (99.5/0.5) as large as 400 % for the lowest tested heat flux. The average heat flux improvement for heat fluxes less than 40 kW/m2 was approximately 105 %, 49 %, and 155 % for the 0.5 %, the 1 %, and the 2 % mass fractions, respectively. The heat flux enhancement for all of the mixtures increased with respect to decreasing heat flux. Due to the good quality of the nanolubricant suspension, the performance of the (99.5/0.5), and the (98/2) nanolubricant mixtures was stable over time.

Citation: NIST TN - 1663

Keywords: additives, boiling, Al2O3, enhanced heat transfer, nanotechnology, refrigerants, refrigerant/lubricant mixtures

NIST TECHNICAL NOTE 1663

Effect of Al$_2$O$_3$ Nanolubricant on R134a Pool Boiling Heat Transfer with Extensive Measurement and Analysis Details

Mark A. Kedzierski

NIST

National Institute of Standards and Technology • U.S. Department of Commerce

NIST TECHNICAL NOTE 1663

Effect of Al$_2$O$_3$ Nanolubricant on R134a Pool Boiling Heat Transfer with Extensive Measurement and Analysis Details

Mark A. Kedzierski

U.S DEPARTMENT OF COMMERCE
National Institute of Standard and Technology
Building Environment Division
Building and Fire Research Laboratory
Gaithersburg, MD 20899-8631

April 2010

U.S. Department of Commerce
Gary F. Locke, Secretary

National Institute of Standards and Technology
Patrick D. Gallagher, Acting Director

Effect of Al$_2$O$_3$ Nanolubricant on R134a Pool Boiling Heat Transfer with Extensive Measurement and Analysis Details

M. A. Kedzierski

National Institute of Standards and Technology
Bldg. 226, Rm B114
Gaithersburg, MD 20899
Phone: (301) 975-5282
Fax: (301) 975-8973

ABSTRACT

This paper quantifies the influence of Al$_2$O$_3$ nanoparticles on the pool boiling performance of R134a/polyolester mixtures on a roughened, horizontal, flat surface. Nanofluids are liquids that contain dispersed nano-size particles. A lubricant based nanofluid (nanolubricant) was made by suspending nominally, 10 nm diameter Al$_2$O$_3$ particles in a synthetic ester to roughly a 1.6 % volume fraction. The nanoparticles enhanced the boiling heat transfer relative to that for R134a/polyolester mixtures without nanoparticles for the three lubricant mass fractions that were tested. The enhancement occurred for the lowest heat fluxes, which gives the opportunity for designing chillers for lower approach temperatures. For the 0.5 % nanolubricant mass fraction, the nanoparticles caused a heat transfer enhancement relative to the heat transfer of pure R134a/polyolester (99.5/0.5) as large as 400 % for the lowest tested heat flux. The average heat flux improvement for heat fluxes less than 40 kW/m^2 was approximately 105 %, 49 %, and 155 % for the 0.5 %, the 1 %, and the 2 % mass fractions, respectively. The heat flux enhancement for all of the mixtures increased with respect to decreasing heat flux. A semi-empirical model was developed to predict the boiling enhancement as cause by the interaction of the nanoparticles with the bubbles. The model suggests that small particle size and large nanoparticle volume fraction improves boiling enhancement. Continued research with nanolubricants and refrigerants are required to further validate the model with measurements with other nanoparticle materials, sizes and concentrations.

Keywords: additives, boiling, Al$_2$O$_3$, enhanced heat transfer, nanotechnology, refrigerants, refrigerant/lubricant mixtures

TABLE OF CONTENTS

LIST OF FIGURES

LIST OF TABLES

INTRODUCTION

In recent years, nanofluids, i.e., liquids with dispersed nano-size particles, have been shown to be a potential means for enhancing the performance of chillers (Liu et al., 2009; Kedzierski, 2009a). A major motivation for improving chiller performance is that energy efficiency is a primary component for net zero energy, high performance green building-design (OSTP 2008, EPA 2008). Chillers that provide air conditioning for buildings consume nearly 13 % of total building electric expenditures (EIA, 2008). Consequently, a cost-effective means for improving the efficiency of chillers would facilitate meeting green building goals.

Lubricant-based nanofluids, i.e., nanolubricants, can facilitate the stability of the nanoparticles in a refrigerant cycle while delivering them to components of the cycle where they can produce the most benefit. Bi et al. (2007a) have shown that nanoparticles in compressor lubricant can improve its performance. Likewise, nanoparticles in the lubricant excess layer that covers evaporator surfaces can interact with bubbles and cause a heat transfer enhancement (Kedzierski, 2009a). The combined effects of nanoparticles on heat transfer and compressor performance was illustrated by Bi et al. (2007b) when they showed that nanolubricants produced energy savings of more than 25 % in domestic refrigerators. The preceding studies suggest that it is worthwhile to investigate the potential benefits of nanolubricants for chillers.

Nanoparticle properties are crucial for determining the performance characteristics of nanolubricants. According to Bobbo et al. (2009), the way in which nanoparticle material, dimension, shape and concentration affect nanolubricant properties is complex and not well understood. Marquis and Chibante (2005) point out that nanoparticle size is more influential in determining thermal conductivity than is the shape of the nanoparticle. Kedzierski (2009a) has shown that the concentration of copper (II) oxide (CuO) nanoparticles may determine whether an enhancement or a degradation in refrigerant/lubricant boiling heat transfer is obtained. The same study also showed that the improvement in thermal conductivity was not the governing factor in determining the magnitude of the enhancement despite CuO having nearly two-orders-of-magnitude greater thermal conductivity than the base lubricant. Peng et al. (2010) and Kedzierski (2009b) have investigated the pool boiling of refrigerant/diamond nanolubricants. Kedzierski (2009b) showed that a good dispersion is necessary to have and maintain an enhancement via particle interaction. Clearly, knowledge of how the properties of nanoparticles influence the heat transfer behavior of nanolubricants must be obtained before the performance can be optimized.

In order to further investigate the influence of nanoparticle properties on refrigerant/lubricant pool boiling, the boiling heat transfer of three R134a/nanolubricant mixtures on a roughened, horizontal, flat (plain), copper surface were measured. A commercial polyolester lubricant (RL68H[1]) with a nominal kinematic viscosity of 72.3 μm^2/s at 313.15 K was the base lubricant that was mixed with nominally 20 nm diameter aluminum oxide (Al_2O_3) nanoparticles. Al_2O_3

---

[1] Certain commercial equipment, instruments, or materials are identified in this paper in order to specify the experimental procedure adequately. Such identification is not intended to imply recommendation or endorsement by the National Institute of Standards and Technology, nor is it intended to imply that the materials or equipment identified are necessarily the best available for the purpose.

nanoparticles have the advantage of a well-established, successful dispersion technology and being relatively inert with respect to lubricated compressor parts.

A manufacturer used a proprietary surfactant at a mass between 15 % and 20 % of the mass of the Al_2O_3 as a dispersant for the RL68H/Al_2O_3 mixture (nanolubricant). The manufacturer made the mixture such that 40 % of the mass was Al_2O_3 particles. The mixture was diluted in-house to a 5.6 % mass fraction of Al_2O_3 by adding neat RL68H and ultrasonically mixing the solution for approximately 24 h. A dynamic light scattering technique suggested that the diameter of most of the particles were approximately 10 nm and that the particles were well dispersed.

The mass faction was chosen so that it matched the nanoparticle mass fraction of the RL68H/copper-oxide study (Kedzierski, 2009a). The RL68H/Al_2O_3 (98.4/1.6)[2] volume fraction mixture, a.k.a. RL68H1AlO, was mixed with pure R134a to obtain three R134a/RL68H1AlO mixtures at nominally 0.5 %, 1 %, and 2 % mass fractions for the boiling tests. In addition, the boiling heat transfer of three R134a/RL68H mixtures (0.5 %, 1 %, and 2 % mass fractions), without nanoparticles, was measured to serve as a baseline for comparison to the RL68H1AlO mixtures.

APPARATUS
Figure 1 shows a schematic of the apparatus that was used to measure the pool boiling data of this study. More specifically, the apparatus was used to measure the liquid saturation temperature (T_s), the average pool-boiling heat flux (q''), and the wall temperature (T_w) of the test surface. The three principal components of the apparatus were the test chamber, the condenser, and the purger. The internal dimensions of the test chamber were 25.4 mm × 257 mm × 1.54 m. The test chamber was charged with approximately 7 kg of refrigerant, giving a liquid height of approximately 80 mm above the test surface. As shown in Fig. 1, the test section was visible through two opposing, flat 150 mm × 200 mm quartz windows. The bottom of the test surface was heated with high velocity (2.5 m/s) water flow. The vapor produced by liquid boiling on the test surface was condensed by the brine-cooled, shell-and-tube condenser and returned as liquid to the pool by gravity. Further details of the test apparatus can be found in Kedzierski (2002) and Kedzierski (2001a).

TEST SURFACE
Figure 2 shows the oxygen-free high-conductivity (OFHC) copper flat test plate used in this study. The test plate was machined out of a single piece of OFHC copper by electric discharge machining (EDM). A tub grinder was used to finish the heat transfer surface of the test plate with a crosshatch pattern. Average roughness measurements were used to estimate the range of average cavity radii for the surface to be between 12 μm and 35 μm. The relative standard uncertainty of the cavity measurements were approximately ± 12 %. Further information on the surface characterization can be found in Kedzierski (2001a).

MEASUREMENTS AND UNCERTAINTIES
The standard uncertainty is the positive square root of the estimated variance. The individual

---

[2] The equivalent mixture is RL68H/CuO (94.4/5.6) in terms of mass.

standard uncertainties are combined to obtain the expanded uncertainty (U), which is calculated from the law of propagation of uncertainty with a coverage factor. All measurement uncertainties are reported at the 95 % confidence level except where specified otherwise. For the sake of brevity, only a summary of the basic measurements and uncertainties is given below. Complete detail on the heat transfer measurement techniques and uncertainties can be found in Kedzierski (2000) and Appendix A, respectively.

All of the copper-constantan thermocouples and the data acquisition system were calibrated against a glass-rod standard platinum resistance thermometer (SPRT) and a reference voltage to a residual standard deviation of 0.005 K. Considering the fluctuations in the saturation temperature during the test and the standard uncertainties in the calibration, the expanded uncertainty of the average saturation temperature was no greater than 0.04 K. Consequently, it is believed that the expanded uncertainty of the temperature measurements was less than 0.1 K.

Twenty 0.5 mm diameter thermocouples were force fitted into the wells of the side of the test plate shown in Fig. 2. The heat flux and the wall temperature were obtained by regressing the measured temperature distribution of the block to the governing two-dimensional conduction equation (Laplace equation). In other words, rather than using the boundary conditions to solve for the interior temperatures, the interior temperatures were used to solve for the boundary conditions following a backward stepwise procedure given in Kedzierski (1995)[3]. The origin of the coordinate system was centered on the surface with respect to the y-direction at the heat transfer surface. Centering the origin in the y-direction reduced the uncertainty of the wall heat flux and temperature calculations by reducing the number of fitted constants involved in these calculations.

Fourier's law and the fitted constants from the Laplace equation were used to calculate the average heat flux (q'') normal to and evaluated at the heat transfer surface based on its projected area. The average wall temperature (T_w) was calculated by integrating the local wall temperature (T). The wall superheat was calculated from T_w and the measured temperature of the saturated liquid (T_s). Considering this, the relative expanded uncertainty in the heat flux ($U_{q''}$) was greatest at the lowest heat fluxes, approaching 10 % of the measurement near 20 kW/m^2. In general, the $U_{q''}$ remained approximately between 3 % and 6 % for heat fluxes greater than 50 kW/m^2. The average random error in the wall superheat (U_{Tw}) remained between 0.06 K and 0.14 K. Plots of $U_{q''}$ and U_{Tw} versus heat flux can be found in Appendix A.

EXPERIMENTAL RESULTS
Heat Transfer
The heat flux was varied approximately between 7 kW/m^2 and 130 kW/m^2 to simulate a range of possible operating conditions for R134a chillers. All pool-boiling measurements were made at 277.6 K saturated conditions. The data were recorded consecutively starting at the largest heat flux and descending in intervals of approximately 4 kW/m^2. The descending heat flux procedure minimized the possibility of any hysteresis effects on the data, which

---

[3] For the record, Table 1 provides functional forms of the Laplace equation that were used in this study in the same way as was done in Kedzierski (1995) and in similar studies by this author.

would have made the data sensitive to the initial operating conditions. Table 2 presents the measured heat flux and wall superheat for all the data of this study. Table 3 gives the number of test days and data points for each fluid. A total of 2225 measurements were made over 46 days.

The mixtures were prepared by charging the test chamber (see Fig. 1) with pure R134a to a known mass. Next, a measured mass of nanolubricant or lubricant was injected with a syringe through a port in the test chamber. The refrigerant/lubricant solution was mixed by flushing pure refrigerant through the same port where the lubricant was injected. All compositions were determined from the masses of the charged components and are given on a mass fraction basis. The maximum uncertainty of the mass fraction measurement is approximately 0.02 %, e.g., the range of a 2.0 % mass fraction is between 1.98 % and 2.02 %. Nominal or target mass compositions are used in the discussion. For example, the "actual" mass composition of the RL68H in the R134a/ RL68H (99.5/0.5) mixture was 0.49 % ± 0.02 %. Likewise, the RL68H mass fractions for R134a/ RL68H (99/1) and the R134a/ RL68H (98/2) mixtures were 1.00 % ± 0.02 % and 2.01 % ± 0.02 %, respectively. Using the same uncertainties, the nanolubricant mass fractions as tested with R134a were 0.50 %, 0.99 %, and 1.99 %.

Figure 3 is a plot of the measured heat flux (q'') versus the measured wall superheat ($T_w - T_s$) for pure R134a pool boiling at a saturation temperature of 277.6 K. These measurements serve as a baseline for comparison to the refrigerant/pure-lubricant measurements. The open triangles represent the measured data while the solid line is a cubic best-fit regression or estimated means of the data. Five days of boiling pure R134a produced 213 measurements over a period of a little over a week. Ten of the 213 R134a/RL68H (99.5/0.5) measurements were removed before fitting because they were identified as "outliers" based on having both high influence and high leverage (Belsley et al., 1980). The data sets for each test fluid presented in this manuscript exhibited a similar number of outliers and were regressed in the same manner. Table 4 gives the constants for the cubic regression of the superheat versus the heat flux for all of the fluids tested here. The residual standard deviation of the regressions - representing the proximity of the data to the mean - are given in Table 5. The dashed lines to either side of the mean represent the lower and upper 95 % simultaneous (multiple-use) confidence intervals for the mean. From the confidence intervals, the expanded uncertainty of the estimated mean wall superheat was, on average, 0.10 K and 0.03 K for superheats less than and greater than 7 K, respectively. Table 6 provides the average magnitude of 95 % multi-use confidence interval for the fitted wall superheat for all of the test data.

The dotted line in Fig 3 illustrates the effect of surface aging for the flat surface of this study. Marto and Lepere (1982) have also observed a surface aging effect where pool boiling is sensitive to the past history of the surface. The dotted line is the mean of measurements for pure R134a that were taken on the same surface approximately 2.5 years prior to the current measurements (Kedzierski and Gong, 2009). The maximum deviation between the measurement means for the two data sets is less than 0.9 K. For a given heat flux, the slope of the newer boiling curve is larger than the one measured 2.5 years prior to the present data. However, the boiling heat transfer performance as degraded over the years for heat fluxes less than 80 kW/m^2.

Figure 4 is a plot of the measured heat flux (q'') versus the measured wall superheat (T_w - T_s) for the refrigerant/pure-lubricant mixtures at a saturation temperature of 277.6 K. The refrigerant/pure-lubricant measurements serve as the baseline for comparison to the refrigerant/nanolubricant pool boiling measurements. Nineteen boiling curves were measured over the span of approximately two months. The open circles, squares, and stars represent the measured heat flux (q'') versus the measured wall superheat (T_w - T_s) at a saturation temperature of 277.6 K for the R134a/RL68H (99.5/0.5), R134a/RL68H (99/1), and R134a/RL68H (98/2) mixtures, respectively. From the 95 % multi-use confidence intervals, the expanded uncertainty of the estimated mean wall superheat was, on average, 0.06 K and 0.05 K for superheats less than and greater than roughly 8 K, respectively.

A general overview of the effect that the variation in the pure lubricant mass fraction has on R134a/lubricant pool boiling can be obtained from Fig. 4. Comparison of the three mean boiling curves shows that the superheats are within approximately 1.2 K of each other for the entire tested heat flux range. For the most part, the superheat for the refrigerant/lubricant mixtures is 0.1 K to 1.8 K greater than that for pure R134a indicating a heat transfer degradation with respect to pure R134a. Kedzierski (2001b) has shown that, in general, degradations associated with increased lubricant mass fractions occur when the concentration induced bubble size reduction, and its accompanying loss of vapor generation per bubble, is not compensated by an increase in site density. Typically, heat transfer degradations have been observed to increase with respect to increasing lubricant mass fraction. The present measurements are inconsistent with this trend in that the heat transfer performance of the (99/1) mixture is better than that of the (99.5/0.5) mixture. Although the measurements are inconsistent with typical refrigerant/lubricant data, they are repeatable within this study. In addition, an inconsistent boiling behavior of the RL68H lubricant was also seen by Kedzierski and Gong (2009) and Kedzierski (2009a), but not with the same relative performances. Comparison of the RL68H boiling measurements of the three studies, shows that, in general, (1) the heat transfer surface performs better as it ages; (2) the boiling performance of the (99.5/0.5) and the (99/1) mixtures are nearly the same or very close ; and (3) the boiling performance of the (98/2) mixture relative to the lesser mass fraction mixtures was better for two of the studies and worse for one of the studies.

A more precise comparison of the R134a/RL68H heat transfer performances relative to pure R134a is given in Fig. 5. Figure 5 plots the ratio of the R134a/RL68H mixture heat flux to the pure R134a heat flux (q''_{PL}/q''_p) versus the pure R134a heat flux (q''_p) at the same wall superheat. Figure 5 illustrates the influence of lubricant mass fraction on the R134a/RL68H boiling curve with solid and dashed lines representing the mean heat flux ratios for each mixture and shaded regions showing the 95 % multi-use confidence level for each mean. Overall, lubricant for all compositions has caused a heat transfer degradation relative to the mean heat transfer of pure R134a for all measured q''_p. For the most part, the mean refrigerant/lubricant pool boiling heat flux resides between roughly 99 % and 48 % of that of the pure refrigerant. The heat flux ratio (q''_{PL}/q''_p) for each mixture is roughly constant between 10 kW/m^2 and 110 kW/m^2: 0.68, 0.92, and 0.52 for the R134a/RL68H (99.5/0.5), the R134a/RL68H (99/1), and the R134a/RL68H (98/2) mixtures, respectively. The smallest heat transfer degradation for each mixture is shown in Fig. 5 and occurs at approximately

20 kW/m^2: 0.76 ± 0.25, 0.99 ± 0.03, and 0.62 ± 0.10 for the R134a/RL68H (99.5/0.5), the R134a/RL68H (99/1), and the R134a/RL68H (98/2) mixtures, respectively.

Figure 6 shows the measured heat flux (q'') versus the measured wall superheat (T_w - T_s) for three mixtures of R134a and the nanolubricant RL68H1AlO at a saturation temperature of 277.6 K. Twenty-two boiling curves were measured over the span of approximately five weeks. The closed circles, squares, and stars represent the measurements for the R134a/ RL68H1AlO (99.5/0.5), R134a/ RL68H1AlO (99/1), and R134a/ RL68H1AlO (98/2) mixtures, respectively. From the 95 % multi-use confidence intervals, the expanded uncertainty of the estimated mean wall superheat was, on average, 0.17 K.

As Fig. 6 shows, a modest variation in the boiling performance is exhibited among the three nanofluids given the variation in composition. For example, the mean boiling curves of the refrigerant/nanolubricant mixtures differ by no more than 1.4 K. For heat fluxes between 40 kW/m^2 and 100 kW/m^2, the difference in superheat of adjacent boiling curves is approximately 0.5 K. In this region, the boiling performance of the R134a/RL68H1AlO mixtures is ranked from best to worst as follows: (98/2), (99.5/0.5), and (99/1). As the nanolubricant mass fraction increases, so does the between-run variation in the measurements. In other words, as the nanolubricant mass fraction increases from 0.5 % to 1 %, the boiling performance degrades. Conversely, as the nanolubricant mass fraction increases from 1 % to 2 %, the boiling performance improves.

Figure 7 summarizes the influence of Al$_2$O$_3$ nanoparticles on R134a/RL68H boiling heat transfer. The figure plots the ratio of the R134a/RL68H1AlO heat flux to the R134a/RL68H heat flux (q''_{Al}/q''_{PL}) versus the R134a/RL68H mixture heat flux (q''_{PL}) at the same wall superheat. The three different compositions are represented by three solid lines where each R134a/nanolubricant mixture is compared to the R134a/pure-lubricant mixture at the same mass fraction. A heat transfer enhancement exists where the heat flux ratio is greater than one and the 95 % simultaneous confidence intervals (depicted by the shaded regions) do not include the value one. Figure 7 shows that all of the R134a/RL68H1AlO mixtures exhibit a boiling heat transfer enhancement over that of the R134a/RL68H mixtures (without nanoparticles), for heat fluxes less than 40 kW/m^2. The average heat flux ratio for heat fluxes less than 40 kW/m^2 is approximately 2.05, 1.49, and 2.55 for the 0.5 %, the 1 %, and the 2 % mass fractions, respectively. Overall, the average heat flux ratio for the (99.5/0.5) mixture, the (99/1) mixture, and the (99/2) mixture from approximately 5 kW/m^2 to 115 kW/m^2 was 1.37, 1.00, and 1.73, respectively. The R134a/RL68H1AlO (99.5/0.5) mixture shows a maximum heat flux ratio of 5.0 ± 0.05 at approximately 2.5 kW/m^2. The R134a/RL68H1AlO (99/1) mixture shows a maximum heat flux ratio of 2.8 ± 0.03 at approximately 5 kW/m^2. The R134a/RL68H1AlO (98/2) mixture shows a maximum heat flux ratio of 3.5 ± 0.05 at approximately 7 kW/m^2. The heat flux ratio for all of the mixtures decreased with respect to increasing heat flux. Aluminum oxide nanoparticles provided the most favorable benefit to the 2 % mass fraction mixture.

Comparisons of the modified Rohsenow (1952) model of Peng et al. (2010) for nucleate pool boiling of refrigerant/nanolubricant mixtures to the present data were attempted. Peng et al.'s (2010) modified fluid-surface constant and their modified exponent on the Prandtl number combined to produce a product that was roughly two-orders of magnitude smaller than that of the original Rohsenow (1952) model. As a result, the Peng et al. (2010) model did not produce realistic results that could be used for comparison, even when the fluid properties of the refrigerant/nanolubricant mixture were used in the modified Rohsenow (1952) model as required. Errors in the presentation of the regressed model coefficients in their paper caused the unrealistic predictions (Peng and Ding, 2010).

Particle Size

The size of the Al_2O_3 nanoparticles in the nanolubricant was measured with a Dynamic Light Scattering (DLS) technique using a 633 nm wavelength laser and a sieving technique using a syringe filter. An index of refraction of 1.67 for Al_2O_3 was used in the Brownian motion based calculation that was done internally by the DLS instrument for the particle size. The uncertainty of the packaged DLS instrumentation was confirmed with a NIST traceable 60 nm nanofluid standard. The measured diameter of the standard with the DLS system was within 5 nm of range of uncertainty of the standard.

Approximately 0.01 g of the nanolubricant was mixed with approximately 3 g of toluene. Samples that were taken from this prepared mixture and pushed through a 100 nm syringe filter and analyzed with the DLS system produced an average diameter of 10.1 nm ± 1.3 nm. Unfiltered samples produced diameters of nominally 30 nm with approximately the same intensity as the filtered sample. Consequently, the nanoparticles were well dispersed in the nanolubricant with mostly discrete, nominally 10 nm Al_2O_3 nanoparticles. Figure 8 shows a Transmission Electron Microscopy (TEM) image of the nanoparticles as taken by Sarkas (2009). The image confirms the good dispersion and shows that the particles are spherical with most of them having diameters of approximately 10 nm or less and a few with diameters close to 50 nm.

MODEL DEVELOPMENT

This section outlines the development of a model for the prediction of the enhancement of refrigerant/lubricant pool boiling as caused by nanoparticles. It is assumed that the transfer of momentum from the nanoparticles to the bubbles is responsible for the boiling heat transfer enhancement and that the interaction is confined to the lubricant excess layer that resides on the boiling surface. The analysis is focused on the premise that bubbles grow through nanoparticles that are suspended in the lubricant excess layer by Brownian motion. The model assumes that a good dispersion exists such that the nanoparticles remain discrete with no agglomeration.

The first part of the model is based on conservation of momentum for nanoparticles impacting a single bubble. Because the velocity of a bubble (u_b) is orders of magnitude larger than the velocity of the nanoparticles (u_{np}), along with the fact that the bubble's velocity is primarily normal to and away from the boiling surface, only the scalar velocities are considered. The sum of the nanoparticles and the bubble momentum is conserved before (subscript "i") and after impact (subscript "f"):

$$\frac{N_{np}}{N_b} M_{np} u_{np_i} + M_{b_i} u_{b_i} = \frac{N_{np}}{N_b} M_{np} u_{np_f} + M_{b_f} u_{b_f} \qquad (1)$$

13

Here, M_{np} is the mass of a single nanoparticle, M_b is the mass of a single bubble, and N_{np}/N_b is the ratio of the number of nanoparticles per bubble in the lubricant excess layer.

It is assumed that the nanoparticles do not change the bubble velocity, but rather, perform surface work to increase the bubble mass upon impact by $\Delta M_b (= M_{b_f} - M_{b_i})$ in exchange for a reduction in nanoparticle velocity ($\Delta u_{np} = u_{np_i} - u_{np_f}$). The preceding assumptions can be used to simplify eq. (1) and solve for the increase in bubble mass as:

$$\Delta M_b = \frac{N_{np}}{N_b} \frac{M_{np}}{u_b} \Delta u_{np} = \frac{N_{np}}{N_b} \frac{\pi D_{np}^3 \rho_{np}}{6 u_b} \Delta u_{np} \tag{2}$$

The rightmost side of eq. (2) was obtained by assuming that the mass of the nanoparticle can be calculated with a spherical volume of diameter D_{np} and a uniform nanoparticle mass density, ρ_{np}.

The increase in the bubble mass can also be represented by the difference in the heat flux for with nanoparticles ($q_{np}^{''}$) and without nanoparticles ($q_{PL}^{''}$):

$$q_{np}^{''} - q_{PL}^{''} = f_b \Delta M_b h_{fg} n \tag{3}$$

where f_b is the bubble frequency, h_{fg} is the latent heat of vaporization of the refrigerant and n is the bubble site density.

The ratio of the enhanced heat flux with nanoparticles to boiling without nanoparticles ($q_{np}^{''} / q_{PL}^{''}$) can be obtained from eq. (3) by dividing through by the heat flux for refrigerant/lubricant boiling ($q_{PL}^{''}$) and assuming that nanoparticles change neither the bubble frequency nor the site density:

$$\frac{q_{np}^{''}}{q_{PL}^{''}} = 1 + \frac{\Delta M_b}{M_b} = 1 + \frac{6 \Delta M_b}{\pi D_b^3 \rho_v} \tag{4}$$

As the rightmost side of eq. (4) shows, the mass of the bubble as it would have been without nanoparticles (M_b) can by calculated by assuming that the bubble is a sphere of diameter D_b with mass density ρ_v (refrigerant vapor). The bubble diameter can be calculated from the refrigerant/lubricant pool-boiling model of Kedzierski (2003):

$$\frac{D_b}{2} = \frac{0.75 l_a \rho_L (1 - x_b)}{x_b \rho_v} = \frac{18.75 \text{Å} \rho_L (1 - x_b)}{x_b \rho_v} \tag{5}$$

where x_b is the mass fraction of the lubricant in the bulk mixture, and ρ_L is the mass density of the pure lubricant. The l_a represents the thickness of the lubricant excess layer that is removed by a departing bubble and is taken as two monolayers, which is approximately 25 Å for lubricant. Equation (5) is valid for $x_b > 0$.

An expression for N_{np}/N_b can be obtained by combining the relationship between N_{np} per volume ($= 6\phi / \pi D_{np}^3$) and an expression given by Gaertner (1965) relating the boiling heat flux to the bubble site density:

$$\frac{N_{np}}{N_b} = \frac{6\phi}{\pi D_{np}^3 K (q_{PL}'')^{3/2}} \qquad (6)$$

Gaertner (1965) notes that the leading constant in the relationship between site density and heat flux is a strong function of the fluid/surface combination while the exponent on the heat flux (3/2) is not. Here, the constant K in eq. (6) above is not the same as used by Gaertner (1965) because the site density has been replaced, for the purpose of the present model development, by the number of bubbles per volume in the lubricant excess layer. As a result, the K given in eq. (6) accounts for the required conversion from bubbles per area to bubbles per volume while assuming that the two parameters are proportional.

Because the boiling analysis is confined to the lubricant excess layer, the bubble velocity of concern is more closely related to the velocity during growth rather than that which would be predicted by the terminal velocity using Stokes Law. The velocity of the bubble during growth can be approximated by its diameter divided by the waiting time, which is equivalent to twice the bubble diameter times the bubble frequency. Jakob (1949) has shown that the product $D_b f$ is very close to a constant. For this reason, the bubble velocity is approximated as a constant and combined with the other constants of this analysis.

The only remaining parameter to be determined to solve for the excess vapor mass from eq. (2) is the reduction in the nanoparticle velocity upon collision, Δu_{np}. The change in kinetic energy of the nanoparticle is equal to the work that nanoparticle does to increase the surface area of the bubble:

$$\frac{1}{2} M_{np} u_{np_f}^2 - \frac{1}{2} M_{np} u_{np_i}^2 = 4\pi\sigma (r_{b_f'}^2 - r_{b_i'}^2) \qquad (7)$$

The primes on the bubble radius subscripts denote that the bubble radius change $(\Delta r_b' = r_{b_f'} - r_{b_i'})$ is only that due to a single nanoparticle and not associated with the total ΔM_b.

Combining like terms of eq. (7), expanding the squared terms to obtain expressions for average values for the bubble diameter and the nanoparticle velocity, and solving for the change in nanoparticle velocity yields:

$$\Delta u_{np} = \frac{24\sigma D_b \Delta r_b'}{D_{np}^3 \rho_{np} u_{np}} = \frac{F\sigma D_b v_L}{D_{np}^4 \rho_{np} \left(\dfrac{\rho_{np}}{\rho_L} - 1 \right) g} \qquad (8)$$

15

where ν_L is the kinematic viscosity of the pure lubricant. The rightmost part of eq. (8) was obtained by assuming that the incremental change in bubble radius is proportional to the diameter of the nanoparticle and that the average nanoparticle velocity can be calculated using the Stokes velocity for Brownian motion (Lamb, 1945). The constant F incorporates the proportional constant for the bubble radius change and those from the Stokes velocity and eq. (8).

Substitution of eqs. (2), (5), (6), and (8) into eq. (4) along with the approximation for the bubble velocity gives:

$$\frac{q_{np}^{"}}{q_{PL}^{"}} = 1 + \frac{3.45 \times 10^{-9} [s] \phi \sigma \nu_L \rho_v x_b^2}{D_{np}^4 \left(q_n^{"}\right)^{3/2} \rho_L \left(\rho_{np} - \rho_L\right) g (1 - x_b)^2} \tag{9}$$

where $q_n^{"}$ is equal to $q_{PL}^{"}$ normalized by $1\ \mathrm{Wm^{-2}}$. In addition, the constants in eqs. (2), (5), (6), and (8) were combined into a single constant and its value (3.45×10^{-9}) was obtained from a least squares regression with the data presented in Fig. 7.

Equation 9 does not model the effects of the nanolubricant viscosity and the nanolubricant thermal conductivity on the boiling performance. The Al_2O_3 nanoparticles caused a 13 % viscosity increase and a 4.5 % increase in thermal conductivity, see Appendix B and Appendix C, respectively. For the present data set, the Kedzierski (2003) model was used to show that the nanolubricant property effects caused approximately a 5 % increase in the heat transfer. This order-of-magnitude change should be nearly true for all dilute, well-dispersed nanolubricants. If this were not the case, the enhancement ratio obtained from the Kedzierski (2003) model must be added to eq. (9) to account for the enhancement due to nanoparticle property effects. However, in order to simplify the model for dilute solutions, the small property effect was incorporated into the fixed constant of eq. (9).

DISCUSSION
Figure 9 compares the predicted to the measured enhancement ratios for the three R134a/RL68H1AlO mixtures as a function of $q_{PL}^{"}$ using dashed lines and solid lines, respectively. Equation 9 was used to generate the predictions shown in Fig. 9. The properties of the refrigerant and the pure lubricant were evaluated at the test temperature 277.6 K. The volume fraction, diameter, and mass density of the nanoparticle used was 0.0156, 10 nm, and 3600 kg·m^{-3}, respectively. The mass density of the pure lubricant used was 990 kg·m^{-3}.

The shaded regions of Fig. 9 highlight the difference between the predictions and the measurements. For heat fluxes greater than 20 kW/m^2, the (99.5/0.5), the (99/1), and the (98/2) mixture were underpredicted on average by approximately 25 %, 0.2 %, and 6 %, respectively. For heat fluxes between 5 kW/m^2 and 20 kW/m^2, the heat flux ratio for the (99.5/0.5) mixture was underpredicted, on average, by approximately 43 %. For the same heat flux region, the (99/1) and the (98/2) mixtures were overpredicted, on average, by approximately 14 %, and 76 %, respectively. Much of the overprediction of the heat flux ratio for the (99/1) and the (98/2) mixtures occurs for heat fluxes less than 10 kW/m^2 where

16

the slope of the heat flux ratio is large. For example, when heat fluxes that are less than 10 kW/m^2 are excluded, the predictions are within 1.5 % and 41 % for the (99/1) and the (98/2) mixtures, respectively, for heat fluxes between 10 kW/m^2 and 20 kW/m^2.

As can be seen for the preceding discussion, the most favorable agreement between measurement and predictions is exhibited for the (98/2) and the (99/1) mixtures. For this case, the model and the measurements both exhibit larger enhancements for increasing nanolubricant mass fractions for a fixed heat flux. Part of the reason that the measured heat flux ratio for the (99.5/0.5) mixture is not predicted as well as it is for the larger mass fraction mixtures is that it does not follow the trend of increasing heat flux ratios with increasing nanolubricant mass fraction. One plausible reason for this is that the average size of the nanoparticles in the (99.5/0.5) mixture may actually be smaller than the nanoparticles that are in the larger mass fraction mixtures. For example, if the average diameter of the nanoparticles in the (99.5/0.5) mixture is 6 nm rather than 10 nm, the model predictions for the (99.5/0.5) mixture heat flux ratio improves significantly: on average the predictions are within 3 % of the (99.5/0.5) mixture heat flux ratio measurements for heat fluxes larger than 10 kW/m^2. As the number of nanoparticles increases in the tested sample mixture, i.e., as the nanolubricant mass fraction increases, the average size of the nanoparticle for the sample population approaches the average size of the nanolubricant batch containing the entire population as it was manufactured. In other words, the average nanoparticle size for smaller samples and smaller mass fractions has a greater potential for deviation from the average nanoparticle size as it exists in the batch than larger samples and larger mass fractions.

Future research and modeling work is needed to verify the predicted trends at dilute mass nanolubricant fractions and to correctly extrapolate the trends at large nanolubricant mass fractions. It may not be necessary to predict the boiling performance of refrigerant/nanolubricant mixture at the extreme limits of nanoparticle volume fraction, nanolubricant mass fraction or nanoparticle diameter. However, it would be useful to be able to optimize these parameters to maximize boiling performance. As the model stands, the maximum performance is approached for volume fraction and mass fractions nearing unity, and forever decreasing nanoparticle size. It is surmised that the current model is valid for the range of parameters investigated here and not much beyond this range. This may in fact be the only range of parameters that is applicable to the refrigeration and air-conditioning industry. On the other hand, opportunities may be realized with a model that can be used to direct the user to the nanoparticle size, and volume fraction that maximizes heat transfer. Further investigation in this area and with other nanoparticle materials may lead to an improved theory that can be used to develop nanolubricants that improve boiling heat transfer for the benefit of the refrigeration and air-conditioning industry. The potential benefit may be significant considering that the boiling enhancement occurred for the lowest heat fluxes, which gives the opportunity for designing chillers for lower approach temperatures.

CONCLUSIONS

The effect of Al_2O_3 nanoparticles on the boiling performance of R134a/polyolester mixtures on a roughened, horizontal flat surface was investigated. A nanolubricant containing roughly 10 nm diameter Al_2O_3 nanoparticles at 1.6 % volume fraction with a polyolester lubricant was mixed with R134a at three different mass fractions. In general, the Al_2O_3 nanoparticles caused a heat transfer enhancement relative to the heat transfer of pure R134a/polyolester for all three of lubricant mass fractions. The heat flux enhancement for all of the mixtures increased with respect to decreasing heat flux. The enhancement occurred for the lowest heat fluxes, which gives the opportunity for designing chillers for lower approach temperatures. Aluminum oxide nanoparticles provided the most favorable benefit to the largest lubricant mass fraction that was tested, i.e., 2 %. The average heat flux improvement for heat fluxes less than 40 kW/m^2 was approximately 105 %, 49 %, and 155 % for the 0.5 %, the 1 %, and the 2 % mass fractions, respectively.

A semi-empirical model was development to predict the enhancement of refrigerant/lubricant pool boiling as caused by nanoparticles. It was assumed that the transfer of momentum from the nanoparticles to the bubbles is responsible for the boiling heat transfer enhancement. For heat fluxes greater than 20 kW/m^2, the model underpredicted the (99.5/0.5), the (99/1), and the (98/2) mixtures on average by approximately 25 %, 0.2 %, and 6 %, respectively. It was speculated that the poorer prediction for the (99.5/0.5) mixture may have been due to the average size of the nanoparticles in the (99.5/0.5) mixture being smaller than the nanoparticles that are in the larger mass fraction mixtures. The model predicts that the maximum performance is approached for volume fraction and mass fractions nearing unity, and forever decreasing nanoparticle size. Future research is required to validate the model beyond the range of parameters investigated here.

ACKNOWLEDGEMENTS
The National Institute of Standards and Technology (NIST) funded this work. Thanks go to the following NIST personnel for their constructive criticism of the first draft of the manuscript: Mr. S. Nabinger, and Dr. P. Domanski. Thanks go to Dr. I. Shinder of NIST for his constructive criticism of the second draft of the manuscript. Furthermore, the author extends appreciation to Mr. W. Guthrie and Mr. A. Heckert of the NIST Statistical Engineering Division for their consultations on the uncertainty analysis. Boiling heat transfer measurements were taken by Mr. David Wilmering of KT Consulting at the NIST laboratory. The RL68H (EMKARATE RL 68H) was donated by Dr. K. Lilje of CPI Engineering Services, Inc. The RL68H1AlO was manufactured by Nanophase Technologies with an aluminum oxide and dispersant in RL68H especially for NIST.

NOMENCLATURE

English Symbols

A_n regression constant in Table 4 n=0,1,2,3

D diameter, m

F constant in eq. (8)

f_b bubble frequency, s^{-1}

g   gravitational acceleration, $m \cdot s^{-2}$

h_{fg} latent heat of vaporization, $kJ \cdot kg^{-1}$

K constant in eq. (6)

k thermal conductivity, $W \cdot K^{-1} \cdot m^{-1}$

l_a lubricant excess layer removed by bubble, m

M mass, kg

N the number of

n bubble site density, m^{-2}

q'' average wall heat flux, $W \cdot m^{-2}$

q_n'' $= q_{PL}'' /1W \cdot m^{-2}$

r_b bubble radius, m

T temperature, K

T_r $= T/273.15$ K

T_w temperature at roughened surface, K

U expanded uncertainty

u velocity, $m \cdot s^{-1}$

X model terms given in Table 2

x_b bulk lubricant mass fraction

Greek symbols

ΔM_b bubble mass increase ($M_{b_f} - M_{b_i}$), kg

$\Delta' r_b$ fractional bubble radius change, m

ΔT_s wall superheat: T_w - T_s, K

Δu_{np} nanoparticle velocity reduction ($u_{np_i} - u_{np_f}$), $m \cdot s^{-1}$

ν kinematic viscosity, $m^2 \cdot s^{-1}$

σ surface tension of refrigerant, $N \cdot m^{-1}$

ρ density, $kg \cdot m^{-3}$

ϕ nanoparticle volume fraction

English Subscripts

Al   R134a/RL68H1AlO mixture

b   bubble

f   after impact

i   before impact

L   pure lubricant without nanoparticles

nL   nanolubricant

np   nanoparticle

p   pure R134a

PL   refrigerant/pure lubricant (R134a/RL68H) mixture

q'' heat flux

s   saturated state

T_w wall temperature

v   refrigerant vapor

REFERENCES

Belsley, D. A., Kuh, E., and Welsch, R. E., 1980, <u>Regression Diagnostics: Identifying Influential Data and Sources of Collinearity</u>, New York: Wiley.

Bi, S., Shi, L., and Zhang, L., 2007a, "Performance Study of a Domestic Refrigerator Using R134a/Mineral Oil/Nano-TiO$_2$ as Working Fluid," *Proceedings of International Conference of Refrigeration,* Beijing, ICRO7-B2-346.

Bi, S., Shi, L., and Zhang, L., 2007b, "Application of Nanoparticles in Domestic Refrigerators," <u>Applied Thermal Engineering</u>, Vol. 28, pp. 1834-1843.

Bobbo, S., Fedele, L., Fabrizio, M., Barison, S., Battiston, S., Pagura C., 2009, "Influence of Nanoparticles Dispersion in POE Oils on Lubricity and R134a Solubility," *Proceedings of 3rd IIR Conference on Thermophysical Properties and Transport Processes of Refrigerants*, Boulder CO., paper IIR-176.

Einstein, A., 1956, Investigations on the Theory of the Brownian Movement, Dover, New York.

Energy Information Administration (EIA), 2008, "2003 Commercial Buildings Energy Consumption Survey: Consumption and Expenditures Tables," http://www.eia.doe.gov/emeu/cbecs/cbecs2003/detailed_tables_2003/2003set15/2003pdf/c13a.pdf (January 2009).

Environmental Protection Agency (EPA), 2008, "EPA Green Building Strategy," EPA-100-F-08-073.

Gaertner, R. F., 1965, "Photographic Study of Nucleate Pool Boiling on a Horizontal Surface," <u>Journal of Heat Transfer</u>, 87C, 17-29.

Jakob, N, 1949, <u>Heat Transfer</u>, Vol. 1, Wiley, New York, p. 644.

Kedzierski, M. A., 2009a, "Effect of CuO Nanoparticle Concentration on R134a/Lubricant Pool-Boiling Heat Transfer," <u>ASME J. Heat Transfer</u>, Vol. 131, No. 4, 043205.

Kedzierski, M. A., 2009b, "Effect of Diamond Nanolubricant on R134a Pool Boiling Heat Transfer," Proceedings of MNHMT09 2nd ASME Micro/Nanoscale Heat & Mass Transfer International Conference, Shanghai, China, paper 18032.

Kedzierski, M. A., 2009c, "Viscosity and Density of CuO Nanolubricant," Proceedings of 3rd IIR Conference on Thermophysical Properties and Transport Processes of Refrigerants, Boulder CO., paper IIR-177.

Kedzierski, M. A., and Gong, M., 2009, "Effect of CuO Nanolubricant on R134a Pool Boiling Heat Transfer with Extensive Measurement and Analysis Details," <u>Int. J. Refrigeration</u>, Vol. 25, pp. 1110-1122.

Kedzierski, M. A., 2003, "A Semi-Theoretical Model for Predicting R123/Lubricant Mixture Pool Boiling Heat Transfer," Int. J. Refrigeration, Vol. 26, pp. 337-348.

Kedzierski, M. A., 2002, "Use of Fluorescence to Measure the Lubricant Excess Surface Density During Pool Boiling," Int. J. Refrigeration, Vol. 25, pp. 1110-1122.

Kedzierski, M. A., 2001a, "Use of Fluorescence to Measure the Lubricant Excess Surface Density During Pool Boiling," NISTIR 6727, U.S. Department of Commerce, Washington, D.C.

Kedzierski, M. A., 2001b, "The Effect of Lubricant Concentration, Miscibility and Viscosity on R134a Pool Boiling" Int. J. Refrigeration, Vol. 24, No. 4., pp. 348-366.

Kedzierski, M. A., 2000, "Enhancement of R123 Pool Boiling by the Addition of Hydrocarbons," Int. J. Refrigeration, Vol. 23, pp. 89-100.

Kedzierski, M. A., 1995, "Calorimetric and Visual Measurements of R123 Pool Boiling on Four Enhanced Surfaces," NISTIR 5732, U.S. Department of Commerce, Washington.

Lamb, H., 1945, Hydrodynamics, 6th Ed., Dover, New York, p. 599.

Lee, S., Choi, S., Li, S, and Eastman, J., A., 1999, "Measuring Thermal Conductivity of Fluids Containing Oxide Nanoparticles," ASME J. Heat Transfer, Vol. 121, pp. 280-289.

Liu, M. S., Lin, M.C.C., Liaw, J.S., Hu, R. Wang, C.C., 2009, "Performance Augmentation of a Water Chiller System Using Nanofluids," Proceedings of ASHRAE Winter Conference, Chicago, CH-09-058.

Marquis, F. D. S., and Chibante, L. P. F., 2005, "Improving the Heat Transfer of Nanofluids and Nanolubricants with Carbon Nanotubes," J. Minerals, Metals and Materials Society, Vol. 57, No. 12, pp. 32-43.

Marto, P. J. and Lepere, V. J., 1982, "Pool Boiling Heat Transfer From Enhanced Surfaces to Dielectric Fluids," ASME Journal of Heat Transfer, Vol. 104, pp. 292-299.

Office of Science and Technology Policy (OSTP), 2008, "Federal Research and Development Agenda for Net-Zero Energy, High-Performance Green Buildings," National Science and Technology Council Committee on Technology, http://www.ostp.gov/galleries/ NSTC%20Reports/FederalRDAgendaforNetZeroEnergyHighPerformanceGreenBuildings.pd f (March 2009).

Peng, H., Ding, G., Hu, H., Jiang, W., Zhuang, D., and Wang, K., 2010, "Nucleate Pool Boiling Heat Transfer Characteristics of Refrigerant/Oil Mixture with Diamond Nanoparticles," Int. J. Refrigeration, Vol. 33, pp. 347-358.

Peng, H., and Ding, G., 2010, Private Communications, Institute of Refrigeration and Cryogenics, Shanghai Jiaotong University, Shanghai, China.

Roder, H. M., Perkins, R. A., Laesecke, A., and Nieto de Castro, C. A., 2000, "Absolute Steady-State Thermal Conductivity Measurements by Use of a Transient Hot-Wire System," J. Res. Natl. Inst. of Stand. and Technol., Vol. 105, No. 2, pp. 221-253.

Rohsenow, W. M., 1952, "A Method of Correlating Heat Transfer Data for Surface Boiling of Liquids," Trans. ASME, Vol. 74, pp. 969-975.

Sibley, L.B., Allen, C.M., Zielenbach, W.J., Peterson, C.L., Goldthwaite, W.H., 1958, "A Study Of Refractory Materials For Seal And Bearing Applications In Aircraft Accessory Units And Rocket Motors," WADC-TR-58-299, 1- 52, AD 203 787.

Sarkas, H., 2009, Private Communications, Nanophase Technologies Corporation, Romeoville, IL.

Vadasz, P., 2006, "Heat Conduction in Nanofluid Suspensions," ASME J. Heat Transfer, Vol. 128, pp. 465-477.

Wasp, F. J., 1977, "Solid-Liquid Flow Slurry Pipeline Transportation, Trans., Tech. Pub., Berlin.

Table 1 Conduction model choice

X_0= constant (all models)	X_1= x	X_2= y	X_3= xy

$$X_4 = x^2 - y^2$$

$$X_5 = y(3x^2 - y^2) \qquad X_6 = x(3y^2 - x^2) \qquad X_7 = x^4 + y^4 - 6(x^2)y^2$$

$$X_8 = yx^3 - xy^3$$

Fluid	Most frequent models
Pure R134a (file: R134a8b.dat)	X_1, X_3, X_4 (126 of 213) 59 % X_1, X_4, X_6 (35 of 213) 16 % X_1, X_4, X_5 (20 of 213) 9 %
R134a/RL68H (99.5/0.5) (file: RL6858b.dat)	X_1, X_2, X_4, X_6 (73 of 203) 36 % X_1, X_3, X_4 (45 of 203) 22 % X_1, X_4, X_5, X_6 (29 of 203) 14 % X_1, X_4, X_5 (23 of 203) 11 %
R134a/RL68H (99/1) (file: RL6818b.dat)	X_1, X_3, X_4 (177 of 552) 32 % X_1, X_2 (44 of 552) 8 % X_1, X_2, X_4 (43 of 552) 8 %
R134a/RL68H (98/2) (file: RL6828b.dat)	X_1, X_3, X_4 (118 of 231) 51 % X_1, X_2, X_4 (60 of 231) 26 % X_1, X_2 (47 of 231) 2 % X_1, X_3, X_4, X_5 (11 of 231) 5 %
R134a/RL681AlO (99.5/0.5) (file: Al205.dat)	X_1, X_5 (198 of 360) 55 % X_1, X_3, X_5 (53 of 360) 15 % X_1, X_2, X_5 (27 of 360) 8 % X_1, X_2, X_3 (25 of 360) 7 %
R134a/RL681AlO (99/1) (file: Al201.dat)	X_1, X_5 (93 of 327) 28 % X_1, X_2 (86 of 327) 26 % X_1, X_3, X_5 (31 of 327) 9 % X_1, X_3, X_5, X_8 (28 of 327) 8 %
R134a/RL681AlO (98/2) (file: Al202.dat)	X_1, X_5 (99 of 339) 29 % X_1, X_2, X_3 (90 of 339) 27 % X_1, X_2, X_3, X_5 (70 of 339) 21 % X_1, X_2 (41 of 339) 12 %

Table 2 Pool boiling data

Pure R134a
File: R134a8b.dat

ΔT_s (K)	q'' (W/m^2)
8.24	126286.
8.22	127452.
8.22	129044.
8.19	122753.
8.20	123093.
8.20	124022.
8.15	113672.
8.14	114141.
8.13	114733.
8.05	102171.
8.04	101640.
8.03	101779.
7.92	88744.
7.91	88700.
7.90	88329.
7.81	79924.
7.77	79175.
7.76	79303.
7.63	68218.
7.59	65143.
7.63	66847.
7.35	52400.
7.35	53091.
7.35	53761.
7.14	44074.
7.14	44075.
7.15	44617.
6.84	34842.
6.82	34528.
6.82	34650.
6.49	26706.
6.46	26284.
6.47	26442.
6.00	18642.
5.98	18522.
5.94	18038.
5.30	11134.
5.24	10906.
5.25	11027.
4.05	7610.
3.82	7292.
3.68	7038.
2.48	2454.
8.32	133791.
8.32	132550.
8.30	129060.
8.21	117335.
8.20	118704.
8.19	119777.
8.12	110244.
8.13	111077.
8.14	111867.
8.03	100777.
8.01	100671.
8.00	101328.
7.90	89536.
7.89	89483.

ΔT_s (K)	q'' (W/m^2)
7.89	90145.
7.76	79322.
7.73	78673.
7.71	78820.
7.59	68952.
7.58	68944.
7.57	69184.
7.38	58449.
7.41	59514.
7.37	58024.
7.22	50597.
7.21	51242.
7.22	51847.
6.91	40004.
6.90	39830.
6.94	41610.
6.57	30904.
6.60	31925.
6.54	30379.
6.10	22125.
6.08	21901.
6.06	21524.
5.47	14423.
5.51	14875.
5.42	14082.
4.10	8164.
4.10	8180.
4.13	8348.
2.70	2546.
2.52	2516.
8.29	122217.
8.30	125218.
8.30	126158.
8.28	120799.
8.28	121341.
8.27	122245.
8.23	110585.
8.23	111125.
8.22	111640.
8.14	101445.
8.09	100761.
8.08	102536.
8.02	90993.
8.01	90667.
8.00	91401.
7.88	80403.
7.86	80506.
7.85	81384.
7.72	69893.
7.72	70666.
7.73	70462.
7.55	60809.
7.53	61005.
7.52	60637.
7.25	47660.
7.27	48448.
7.26	48325.
6.92	37008.
6.91	37892.
6.94	38542.
6.63	30428.
6.69	32379.

ΔT_s (K)	q'' (W/m^2)
6.69	32659.
6.26	22604.
6.24	22499.
6.27	22801.
5.69	14878.
5.60	14318.
5.61	14308.
4.25	7841.
4.08	7591.
8.36	130752.
8.36	130366.
8.35	129597.
8.29	116019.
8.31	120370.
8.31	119556.
8.23	110824.
8.22	111124.
8.20	111827.
8.15	100992.
8.15	101763.
8.15	102515.
8.06	91055.
8.05	91420.
8.02	92311.
7.89	79083.
7.89	79453.
7.90	80233.
7.76	68013.
7.74	68167.
7.73	68949.
7.59	59026.
7.56	58358.
7.52	58054.
7.33	49060.
7.32	48712.
7.33	48667.
7.05	38772.
7.01	37605.
6.98	36906.
6.68	28506.
6.70	30171.
6.68	30041.
6.29	22354.
6.34	23853.
6.36	24341.
5.73	14714.
5.69	14303.
5.67	14315.
4.61	8868.
4.43	8405.
4.39	8297.
2.74	2375.
8.28	114214.
8.27	115717.
8.26	117132.
8.23	106090.
8.22	106845.
8.21	107654.
8.14	99113.
8.14	100284.
8.14	101096.
8.05	88681.

ΔT_s (K)	q'' (W/m^2)
8.04	89060.
8.03	89729.
7.94	79928.
7.93	80246.
7.92	81002.
7.77	68238.
7.77	68721.
7.78	69505.
7.58	57323.
7.56	57416.
7.55	57424.
7.31	47873.
7.28	47283.
7.31	48136.
6.99	37745.
7.01	38545.
7.01	38716.
6.69	30431.
6.65	29905.
6.58	28359.
6.24	22824.
6.26	23307.
6.21	21830.
5.62	14684.
5.63	14646.
5.61	14662.
4.80	10175.
4.76	9999.
4.77	10096.
3.63	2937.
3.43	2922.
3.38	2895.

R134a/RL68H (99.5/0.5)
File: RL6858b.dat

ΔT_s (K)	q'' (W/m^2)
8.83	115485.
8.85	116261.
8.87	116929.
8.79	106889.
8.82	106629.
8.83	107034.
8.75	98713.
8.76	98321.
8.74	98044.
8.59	87001.
8.60	87585.
8.60	87632.
8.42	76575.
8.39	76002.
8.39	76083.
8.24	68850.
8.23	68123.
8.24	68401.
7.99	57347.
7.99	57120.
8.00	57099.
7.75	48084.
7.76	48255.
7.80	49486.
7.42	38200.
7.40	38093.
7.46	39419.

7.14	31584.
7.14	31959.
7.18	32768.
6.60	21272.
6.51	20065.
9.10	124981.
9.11	124786.
9.12	124326.
8.99	114004.
8.96	112225.
8.90	108362.
8.86	105910.
8.85	106576.
8.85	107641.
8.67	94705.
8.67	95486.
8.68	96325.
8.51	85790.
8.52	86588.
8.52	87122.
8.34	77458.
8.30	76365.
8.32	77077.
8.09	65614.
8.13	65585.
8.19	67347.
7.95	56905.
7.96	57121.
7.99	57552.
7.65	45630.
7.66	46095.
7.71	47686.
7.42	39118.
7.39	38301.
7.37	37790.
6.98	28187.
6.97	28321.
7.00	29114.
6.63	22252.
6.58	21380.
6.52	20567.
6.05	14550.
6.04	14825.
6.07	15160.
5.44	11179.
5.40	9539.
5.42	9766.
4.09	2839.
3.90	2739.
3.81	2670.
9.02	117920.
9.04	119055.
9.05	119853.
8.93	111197.
8.91	111131.
8.90	111697.
8.78	101676.
8.75	101780.
8.75	102579.
8.60	91899.
8.61	92136.
8.60	92620.
8.39	81040.
8.44	82761.
8.41	81808.

8.24	71316.
8.31	72967.
8.27	71860.
8.08	62227.
8.11	62120.
8.12	62315.
7.93	54327.
7.93	54253.
7.94	54466.
7.62	43322.
7.59	42266.
7.56	41731.
7.30	34326.
7.29	34067.
7.27	33924.
6.96	26982.
6.95	26768.
6.93	26395.
6.43	18705.
6.33	17457.
6.29	16913.
5.79	11995.
5.74	11651.
5.72	11573.
4.81	7754.
4.75	7670.
4.74	7652.
9.10	126702.
9.15	128230.
9.15	129622.
9.00	120910.
9.00	121777.
9.01	122458.
8.81	109689.
8.85	111381.
8.83	111789.
8.64	100421.
8.62	99779.
8.62	99777.
8.49	89950.
8.49	90266.
8.47	89995.
8.35	81375.
8.36	80785.
8.37	80414.
8.20	70117.
8.21	68410.
8.21	67601.
8.09	60981.
8.09	60796.
8.09	60700.
7.90	52808.
7.89	52720.
7.89	52662.
7.60	42640.
7.60	42557.
7.60	42619.
7.29	33602.
7.28	33089.
7.23	32116.
6.84	24336.
6.84	24031.
6.83	23819.
6.47	19103.
6.47	19142.

ΔT_s (K)	q'' (W/m²)
6.50	19216.
5.86	12205.
5.76	11597.
5.77	11625.
4.73	7211.
4.54	6950.
9.13	124541.
9.14	126683.
9.14	128598.
9.02	119824.
9.05	120677.
9.05	122292.
8.89	110498.
8.87	111327.
8.87	111986.
8.66	99740.
8.71	101991.
8.71	102378.
8.50	88985.
8.52	89399.
8.52	89690.
8.34	79153.
8.34	78685.
8.35	78163.
8.25	70159.
8.27	69884.
8.28	69741.
8.11	59591.
8.10	59266.
8.07	58327.
7.83	48314.
7.86	48887.
7.85	48600.
7.66	42579.
7.66	42256.
7.66	42531.
7.32	33039.
7.32	33243.
7.31	33023.
6.98	26075.
6.98	25799.
6.98	25531.
6.57	19372.
6.44	17736.
6.39	17240.
5.59	9855.

R134a/RL68H (99/1)
File: RL6818b.dat

ΔT_s (K)	q'' (W/m²)
8.94	122306.
9.00	123798.
9.01	123714.
8.85	114167.
8.85	113463.
8.85	113443.
8.72	105803.
8.71	105203.
8.74	104882.
8.50	94219.
8.50	93896.
8.50	93653.
8.29	84424.
8.26	83747.
8.26	83874.
8.06	75644.
8.04	74891.
8.06	75710.
7.84	66135.
7.83	66208.
7.84	66619.
7.62	55412.
7.62	55347.
7.65	55929.
7.41	45749.
7.42	45864.
7.42	46005.
7.19	37736.
7.20	38148.
7.21	37926.
6.83	28236.
6.82	27632.
6.83	28213.
6.33	21511.
6.30	21063.
6.25	20473.
5.70	14311.
5.63	13801.
5.59	13595.
4.81	8489.
4.69	8095.
4.69	8100.
2.43	3074.
8.94	122306.
9.00	123798.
9.01	123714.
8.85	114167.
8.85	113463.
8.85	113443.
8.72	105803.
8.71	105203.
8.74	104882.
8.50	94219.
8.50	93896.
8.50	93653.
8.29	84424.
8.26	83747.
8.26	83874.
8.06	75644.
8.04	74891.
8.06	75710.
7.84	66135.
7.83	66208.
7.84	66619.
7.62	55412.
7.62	55347.
7.65	55929.
7.41	45749.
7.42	45864.
7.42	46005.
7.19	37736.
7.20	38148.
7.21	37926.
6.83	28236.
6.82	27632.
6.83	28213.
6.33	21511.
6.30	21063.

ΔT_s (K)	q'' (W/m²)
6.25	20473.
5.70	14311.
5.63	13801.
5.59	13595.
4.81	8489.
4.69	8095.
4.69	8100.
8.73	126555.
8.76	127346.
8.79	130171.
8.64	127366.
8.59	125005.
8.59	126008.
8.47	115993.
8.50	116428.
8.52	116884.
8.40	108527.
8.44	109006.
8.46	109597.
8.30	100204.
8.30	100349.
8.31	100704.
8.08	87796.
8.08	88254.
8.07	88509.
7.89	79659.
7.87	80141.
7.88	80991.
7.66	69862.
7.65	70332.
7.67	71216.
7.46	60950.
7.46	61382.
7.47	62173.
8.73	126555.
8.76	127346.
8.79	130171.
8.64	127366.
8.59	125005.
8.59	126008.
8.47	115993.
8.50	116428.
8.52	116884.
8.40	108527.
8.44	109006.
8.46	109597.
8.30	100204.
8.30	100349.
8.31	100704.
8.08	87796.
8.08	88254.
8.07	88509.
7.89	79659.
7.87	80141.
7.88	80991.
7.66	69862.
7.65	70332.
7.67	71216.
7.46	60950.
7.46	61382.
8.78	130956.
8.81	130802.
8.85	130444.
8.72	120656.
8.72	119223.

8.64	114313.	7.22	46002.	8.56	107260.
8.57	110069.	7.19	44041.	8.57	108257.
8.57	111253.	7.13	42297.	8.38	96334.
8.58	112348.	6.98	37538.	8.38	96941.
8.39	101962.	7.01	37370.	8.40	97647.
8.40	102242.	7.02	37206.	8.18	86555.
8.42	102711.	6.72	30126.	8.19	87394.
8.24	91710.	6.73	30458.	8.17	87646.
8.22	90875.	6.75	30554.	7.94	77753.
8.22	91490.	6.22	22553.	7.94	78317.
8.02	80932.	6.27	20452.	7.92	78845.
8.00	80475.	6.19	21824.	7.75	68625.
7.98	80139.	5.67	15774.	7.72	68973.
7.84	71183.	5.70	16195.	7.72	69426.
7.86	70970.	5.64	15532.	7.50	57744.
7.89	72942.	8.84	128575.	7.50	57461.
7.73	65125.	8.79	125877.	7.51	56849.
7.72	64357.	8.79	124211.	7.25	46143.
7.67	63336.	8.67	114999.	7.28	45902.
7.53	56576.	8.65	113144.	7.26	44722.
7.47	56486.	8.65	111738.	7.01	36517.
7.44	56207.	8.55	105901.	6.99	36074.
7.22	46002.	8.56	107260.	7.00	36967.
7.19	44041.	8.57	108257.	6.57	28416.
7.13	42297.	8.38	96334.	6.65	27514.
6.98	37538.	8.38	96941.	6.71	28305.
7.01	37370.	8.40	97647.	6.26	22286.
7.02	37206.	8.18	86555.	6.27	22284.
6.72	30126.	8.19	87394.	6.27	22417.
6.73	30458.	8.17	87646.	5.59	14315.
6.75	30554.	7.94	77753.	5.61	14350.
6.22	22553.	7.94	78317.	5.60	14246.
6.27	20452.	7.92	78845.	4.74	8638.
6.19	21824.	7.75	68625.	4.68	8412.
5.67	15774.	7.72	68973.	4.64	8270.
5.70	16195.	7.72	69426.	2.63	3337.
5.64	15532.	7.50	57744.	8.88	130635.
3.67	5031.	7.50	57461.	8.91	130701.
8.78	130956.	7.51	56849.	8.90	130102.
8.81	130802.	7.25	46143.	8.78	121489.
8.85	130444.	7.28	45902.	8.77	120524.
8.72	120656.	7.26	44722.	8.80	122323.
8.72	119223.	7.01	36517.	8.68	115268.
8.64	114313.	6.99	36074.	8.71	116541.
8.57	110069.	7.00	36967.	8.73	117925.
8.57	111253.	6.57	28416.	8.50	103794.
8.58	112348.	6.65	27514.	8.50	104688.
8.39	101962.	6.71	28305.	8.49	105114.
8.40	102242.	6.26	22286.	8.28	92499.
8.42	102711.	6.27	22284.	8.28	92616.
8.24	91710.	6.27	22417.	8.30	93653.
8.22	90875.	5.59	14315.	8.17	86066.
8.22	91490.	5.61	14350.	8.14	85418.
8.02	80932.	5.60	14246.	8.10	85411.
8.00	80475.	4.74	8638.	7.86	74085.
7.98	80139.	4.68	8412.	7.83	73134.
7.84	71183.	4.64	8270.	7.80	72955.
7.86	70970.	2.63	3337.	7.62	64719.
7.89	72942.	8.84	128575.	7.60	64262.
7.73	65125.	8.79	125877.	7.62	64437.
7.72	64357.	8.79	124211.	7.47	56078.
7.67	63336.	8.67	114999.	7.45	56070.
7.53	56576.	8.65	113144.	7.46	55871.
7.47	56486.	8.65	111738.	7.24	46008.
7.44	56207.	8.55	105901.	7.22	44460.

7.22	45037.
7.00	37185.
7.03	37410.
7.02	37415.
6.71	28661.
6.71	28212.
6.74	28996.
6.21	21415.
6.19	21176.
6.19	21075.
5.06	10213.
4.97	9811.
4.93	9753.
8.88	130635.
8.91	130701.
8.90	130102.
8.78	121489.
8.77	120524.
8.80	122323.
8.68	115268.
8.71	116541.
8.73	117925.
8.50	103794.
8.50	104688.
8.49	105114.
8.28	92499.
8.28	92616.
8.30	93653.
8.17	86066.
8.14	85418.
8.10	85411.
7.86	74085.
7.83	73134.
7.80	72955.
7.62	64719.
7.60	64262.
7.62	64437.
7.47	56078.
7.45	56070.
7.46	55871.
7.24	46008.
7.22	44460.
7.22	45037.
7.00	37185.
7.03	37410.
7.02	37415.
6.71	28661.
6.71	28212.
6.74	28996.
6.21	21415.
6.19	21176.
6.19	21075.
5.06	10213.
4.97	9811.
4.93	9753.
8.88	130444.
8.88	129612.
8.83	125558.
8.76	121426.
8.82	124626.
8.81	124370.
8.65	114339.
8.65	113943.
8.66	114980.
8.49	104555.

8.48	104410.
8.48	104432.
8.29	93556.
8.30	93484.
8.28	93242.
8.09	82671.
8.08	83231.
8.06	82883.
7.94	78345.
7.84	75212.
7.85	75581.
7.68	66850.
7.64	64986.
7.63	65377.
7.49	56881.
7.49	56814.
7.51	57553.
7.26	46491.
7.24	46411.
7.28	47249.
6.99	36582.
7.00	36589.
7.00	36019.
6.64	27509.
6.62	29282.
6.61	28944.
6.00	18555.
5.95	18017.
5.97	18010.
4.92	9550.
8.88	130444.
8.88	129612.
8.83	125558.
8.76	121426.
8.82	124626.
8.81	124370.
8.65	114339.
8.65	113943.
8.66	114980.
8.49	104555.
8.48	104410.
8.48	104432.
8.29	93556.
8.30	93484.
8.28	93242.
8.09	82671.
8.08	83231.
8.06	82883.
7.94	78345.
7.84	75212.
7.85	75581.
7.68	66850.
7.64	64986.
7.63	65377.
7.49	56881.
7.49	56814.
7.51	57553.
7.26	46491.
7.24	46411.
7.28	47249.
6.99	36582.
7.00	36589.
7.00	36019.
6.64	27509.
6.62	29282.

6.61	28944.
6.00	18555.
5.95	18017.
5.97	18010.
8.84	121800.
8.82	123012.
8.84	124736.
8.71	114869.
8.72	115085.
8.70	115483.
8.55	106045.
8.56	106657.
8.56	107217.
8.39	95962.
8.38	96286.
8.36	96502.
8.16	85782.
8.14	85771.
8.14	86363.
7.92	76090.
7.92	76793.
7.92	77022.
7.74	67247.
7.73	66969.
7.71	66924.
7.53	56032.
7.51	56005.
7.52	55668.
7.29	44918.
7.29	44996.
7.25	44065.
7.04	36229.
7.03	36087.
7.03	35829.
6.70	27197.
6.62	28128.
6.65	28980.
6.32	22520.
6.31	21964.
6.30	21932.
5.62	13792.
5.50	12984.
5.46	12835.
4.20	6250.
8.84	121800.
8.82	123012.
8.84	124736.
8.71	114869.
8.72	115085.
8.70	115483.
8.55	106045.
8.56	106657.
8.56	107217.
8.39	95962.
8.38	96286.
8.36	96502.
8.16	85782.
8.14	85771.
8.14	86363.
7.92	76090.
7.92	76793.
7.92	77022.
7.74	67247.
7.73	66969.
7.71	66924.

ΔT_s (K)	q'' (W/m²)
7.53	56032.
7.51	56005.
7.52	55668.
7.29	44918.
7.29	44996.
7.25	44065.
7.04	36229.
7.03	36087.
7.03	35829.
6.70	27197.
6.62	28128.
6.65	28980.
6.32	22520.
6.31	21964.
6.30	21932.
5.62	13792.
5.50	12984.
5.46	12835.
4.20	6250.

R134a/RL68H (98/2)
File: RL6828b.dat

ΔT_s (K)	q'' (W/m²)
10.15	131687.
10.20	132675.
10.26	133635.
10.26	134107.
10.29	134429.
10.31	134705.
10.32	134967.
10.31	134752.
10.31	134511.
10.30	134498.
10.10	122935.
10.09	121813.
10.00	117891.
9.81	110803.
9.80	112371.
9.81	113248.
9.46	100006.
9.44	100666.
9.44	101291.
9.19	91156.
9.18	91313.
9.15	91852.
8.96	81932.
8.92	81650.
8.88	81649.
8.70	73175.
8.63	73080.
8.61	73330.
8.42	64671.
9.75	112156.
9.85	114830.
9.89	116786.
9.75	108974.
9.82	110837.
9.83	112020.
9.57	100003.
9.56	101264.
9.57	102171.
9.26	90461.
9.22	91443.
9.22	92364.
8.95	82129.
8.93	83006.
8.93	83594.
8.76	72971.
8.71	73019.
8.64	73355.
8.31	63377.
8.25	63616.
8.22	63692.
8.06	54802.
8.06	54443.
8.06	54341.
7.82	45324.
7.81	45795.
7.82	46268.
7.62	36767.
7.61	36134.
9.89	119084.
9.99	119815.
10.03	120779.
9.84	110114.
9.88	110944.
9.90	111980.
9.65	99227.
9.64	97916.
9.61	98250.
9.38	91203.
9.38	91912.
9.38	92880.
9.19	82808.
9.19	83278.
9.14	84574.
8.92	73874.
8.90	73291.
8.87	73963.
8.70	64289.
8.65	64585.
8.62	64478.
8.35	53389.
8.26	53363.
8.21	52211.
8.11	47403.
8.04	46147.
7.93	48818.
7.72	38901.
7.72	38264.
7.72	38064.
7.51	30965.
7.47	31224.
7.47	31278.
7.22	24065.
10.49	132583.
10.50	132791.
10.50	132659.
10.28	121322.
10.28	121165.
10.27	120959.
10.08	111203.
10.05	111055.
10.00	110836.
9.72	99786.
9.68	99671.
9.64	99665.
9.42	91280.
9.39	91439.
9.37	91642.
9.12	80195.
9.06	80172.
9.01	80956.
8.84	71456.
8.79	71789.
8.78	72473.
8.54	62186.
8.45	62568.
8.39	62636.
8.18	52020.
8.18	51697.
8.16	51216.
7.95	44867.
7.92	43924.
9.70	130729.
9.71	130737.
9.74	130654.
9.65	120298.
9.74	117851.
9.85	115682.
9.74	105827.
9.82	108865.
9.85	110120.
9.64	99375.
9.66	100720.
9.64	101391.
9.44	89700.
9.38	90491.
9.34	90748.
9.16	80111.
9.12	80625.
9.10	80737.
8.91	73291.
8.84	73409.
8.77	73409.
8.51	63683.
8.44	63685.
8.43	63957.
8.26	55152.
8.25	55860.
8.26	56513.
8.01	45140.
8.01	44842.
8.03	46222.
7.76	37388.
7.78	38064.
7.78	38304.
7.45	27264.
7.40	26690.
7.40	26705.
6.81	17117.
9.63	117598.
9.64	113278.
9.71	113312.
9.63	105720.
9.79	106116.
9.83	107497.
9.60	93944.
9.63	94895.
9.66	96279.
9.45	86119.
9.43	86783.
9.42	87972.

ΔT_s (K)	q'' (W/m²)
9.20	77420.
9.19	77270.
9.14	77382.
8.85	67593.
8.75	67554.
8.67	67525.
8.40	55697.
8.37	57500.
8.34	58058.
8.19	50659.
8.19	51127.
8.21	52262.
7.93	43109.
7.92	42685.
7.95	44084.
7.66	33727.
7.66	33773.
7.65	33916.
7.30	24733.
7.27	23850.
7.21	23187.
6.66	16331.
6.68	16291.
6.65	16227.
5.40	8626.
9.80	127653.
9.86	127189.
9.94	126473.
9.93	116070.
10.05	113940.
10.10	113901.
9.92	103354.
9.92	103023.
9.92	103269.
9.66	93264.
9.65	93352.
9.68	94925.
9.48	84398.
9.43	84181.
9.39	83715.
9.22	75503.
9.17	74955.
9.11	75131.
8.92	67267.
8.82	67101.
8.74	67878.
8.48	57462.
8.44	58625.
8.42	59836.
8.24	50010.
8.22	50326.
8.21	51098.
8.01	40765.
7.95	41912.
7.95	39695.
7.66	32435.
7.66	32462.
7.65	32438.
7.30	24076.
7.28	23778.
7.24	23525.

R134a/RL681Al0 (99.5/0.5)
File: Al205.dat

ΔT_s (K)	q'' (W/m²)
10.31	128334.
10.30	128773.
10.27	129074.
10.08	121729.
10.09	123752.
10.07	123740.
9.85	115930.
9.84	115943.
9.82	115693.
9.56	106541.
9.54	106615.
9.52	106310.
9.21	96884.
9.25	98022.
9.30	100021.
8.90	88777.
8.95	89708.
8.98	90370.
8.66	80925.
8.72	81120.
8.74	81466.
8.29	70702.
8.30	70460.
8.30	70030.
7.83	60959.
7.68	58181.
7.51	55167.
7.43	54212.
7.18	50612.
7.26	51099.
6.69	43769.
6.75	44393.
6.78	44624.
6.21	36395.
6.22	36661.
6.21	35995.
5.50	27327.
9.83	131857.
9.77	131512.
9.73	131253.
9.47	121446.
9.44	121481.
9.43	122591.
9.22	115759.
9.20	115702.
9.19	115618.
8.86	104795.
8.84	104595.
8.84	104940.
8.57	96442.
8.56	96384.
8.60	96880.
8.23	86206.
8.27	86085.
8.31	86540.
8.00	78607.
8.05	79058.
8.08	79436.
7.65	69473.
7.66	69591.

ΔT_s (K)	q'' (W/m²)
7.70	70011.
7.22	60029.
7.24	59723.
7.25	60161.
6.85	52222.
6.84	51798.
6.85	51617.
6.24	41134.
6.32	42132.
6.36	42677.
5.79	34397.
5.84	35001.
5.86	35164.
5.32	28115.
5.17	26056.
5.14	25799.
4.65	20411.
4.67	20736.
4.66	20519.
3.88	13428.
3.78	13033.
9.53	118996.
9.54	120369.
9.53	121168.
9.31	113003.
9.29	112989.
9.28	113445.
9.00	104402.
9.00	104561.
9.01	105182.
8.73	96625.
8.69	95904.
8.69	96302.
8.34	86838.
8.35	87117.
8.37	87431.
7.95	77373.
7.97	77582.
7.98	77736.
7.63	70168.
7.62	70184.
7.65	70656.
7.15	60752.
7.11	59843.
7.16	59771.
6.77	52449.
6.76	51238.
6.76	50925.
6.23	41964.
6.22	41410.
6.24	41564.
5.81	35206.
5.76	34219.
5.72	33514.
5.03	25265.
5.02	24993.
5.02	25101.
4.45	18867.
4.40	18493.
4.42	18559.
3.66	12239.
3.63	12083.
3.61	11939.
2.66	6549.
2.55	6291.

2.55	6258.	8.40	75173.	5.06	19592.
1.37	2651.	8.40	75179.	5.04	19457.
1.33	2584.	8.41	75442.	5.07	19588.
10.09	123030.	8.02	66960.	4.12	12203.
9.88	117053.	8.05	67228.	4.00	11650.
9.86	117187.	8.04	67315.	3.97	11429.
9.66	110051.	7.58	58247.	2.41	4515.
9.65	110353.	7.57	57885.	2.15	4191.
9.67	111626.	7.60	58144.	10.07	122951.
9.37	101469.	7.08	48901.	10.04	122427.
9.39	102332.	7.08	48697.	9.96	119246.
9.36	101427.	7.09	48750.	9.72	109545.
9.06	92507.	6.58	40942.	9.71	109784.
9.07	92900.	6.55	40233.	9.72	111055.
9.08	93493.	6.39	38068.	9.47	101894.
8.77	85269.	6.09	34139.	9.47	102572.
8.79	85675.	6.08	33981.	9.45	101695.
8.79	85948.	6.14	34570.	9.20	93208.
8.38	75760.	5.39	25361.	9.20	93466.
8.39	75831.	5.49	26260.	9.22	94413.
8.39	75597.	5.51	26490.	8.91	85100.
8.00	67605.	4.79	18970.	8.89	85145.
8.00	67353.	4.72	18535.	8.89	85068.
8.00	67306.	4.71	18403.	8.55	75431.
7.41	55886.	3.65	10496.	8.52	74938.
7.49	56612.	3.54	9905.	8.52	74856.
7.55	57226.	3.50	9788.	8.17	66241.
7.07	48824.	2.52	5072.	8.16	65795.
7.11	49263.	2.25	4816.	8.16	65682.
7.16	49877.	9.92	116989.	7.78	57427.
6.57	41098.	9.92	123869.	7.76	56825.
6.56	41018.	9.93	124432.	7.78	56872.
6.58	41102.	9.70	115009.	7.38	48978.
6.03	33394.	9.67	114535.	7.40	49035.
6.04	33636.	9.67	114913.	7.38	48737.
6.02	33271.	9.43	106032.	6.80	39711.
5.44	25963.	9.43	106446.	6.80	39628.
5.37	25329.	9.44	107003.	6.80	39569.
5.36	25346.	9.15	97003.	6.20	31238.
4.54	17099.	9.16	97341.	6.15	30701.
4.43	16339.	9.17	97746.	6.15	30738.
4.41	16175.	8.88	87388.	5.65	24822.
3.74	11254.	8.92	87699.	5.62	24548.
3.69	11061.	8.95	88036.	5.59	24155.
3.69	10986.	8.66	79380.	4.85	16994.
2.65	5857.	8.65	78838.	4.79	16694.
2.49	5477.	8.65	78832.	4.78	16528.
2.46	5376.	8.32	70072.	3.82	9874.
1.33	2539.	8.31	69844.	3.96	11234.
1.28	2448.	8.31	69543.	4.07	11407.
9.88	115655.	7.85	59739.	2.66	5047.
9.88	116645.	7.85	59462.	2.41	4768.
9.91	118226.	7.84	59200.	2.36	4669.
9.68	110516.	7.44	51532.	1.63	3042.
9.68	111419.	7.36	49943.	1.57	3011.
9.68	112039.	7.34	49431.	10.18	128895.
9.37	101278.	6.96	43267.	10.16	129035.
9.37	102418.	6.96	43365.	10.14	128872.
9.36	102749.	7.04	44170.	9.89	118861.
9.00	91715.	6.36	34496.	9.84	116848.
9.01	91921.	6.47	35630.	9.79	115764.
9.03	92373.	6.52	36097.	9.66	110528.
8.75	84626.	5.86	27957.	9.68	111551.
8.76	84635.	5.79	27216.	9.67	111250.
8.76	84602.	5.78	27107.	9.40	101801.

9.41	102183.
9.43	103020.
9.14	93060.
9.13	92670.
9.13	92913.
8.83	83660.
8.83	83815.
8.84	84146.
8.51	75230.
8.50	74751.
8.51	74816.
8.14	66131.
8.11	65720.
8.12	65530.
7.78	58298.
7.77	57873.
7.78	57737.
7.26	48241.
7.25	47716.
7.24	47238.
6.65	38245.
6.60	37459.
6.61	37221.
6.23	32180.
6.18	31620.
6.23	32025.
5.53	23807.
5.46	23179.
5.46	23098.
4.91	17847.
4.85	17319.
4.85	17269.
3.88	10798.
3.88	10744.
3.87	10631.
2.79	5877.
2.65	5551.
2.57	5305.
1.54	2549.
1.39	2705.

R134a/RL681Al0 (99/1)
File: Al201.dat

ΔT_s (K)	q'' (W/m²)
11.41	113130.
11.43	113216.
11.48	113920.
11.21	106336.
11.24	106665.
11.27	106963.
10.87	98542.
10.88	98429.
10.93	98794.
10.38	87990.
10.33	87824.
10.31	87887.
9.86	80514.
9.84	80505.
9.84	80731.
9.25	73424.
9.22	73433.
9.19	73336.
8.65	66076.

8.61	65654.
8.62	65775.
8.08	57905.
8.01	57126.
7.99	57264.
7.47	50942.
7.43	50357.
7.44	50086.
6.84	42496.
6.81	41519.
6.80	41258.
6.25	34929.
6.24	34502.
6.25	34257.
5.47	27329.
5.48	27195.
5.49	27044.
4.96	18796.
5.03	18380.
5.12	18537.
3.96	10950.
3.90	10959.
3.84	10692.
2.77	6048.
2.65	5767.
2.59	5636.
1.60	2966.
1.50	2954.
12.22	102199.
12.14	102986.
12.06	116850.
11.42	109301.
11.37	109693.
11.33	109828.
10.72	101553.
10.67	101594.
10.64	101757.
10.11	94142.
10.04	93688.
10.01	93697.
9.48	86153.
9.44	85901.
9.42	85917.
8.93	78857.
8.88	78295.
8.64	72284.
8.19	70897.
8.16	70708.
8.14	70673.
7.47	59804.
7.52	59557.
7.59	59895.
7.19	53014.
7.23	51565.
7.34	51366.
6.85	43757.
6.82	42707.
6.84	42401.
6.50	37453.
6.46	36815.
6.47	37050.
5.76	29182.
5.79	28862.
5.77	28530.
5.08	21996.

5.08	22173.
5.12	22243.
3.94	13266.
3.83	12784.
3.82	12705.
2.30	4938.
2.12	4504.
9.89	119414.
9.89	120436.
9.90	122201.
9.52	114258.
9.50	114671.
9.42	113465.
9.02	104448.
8.99	104535.
9.00	105196.
8.61	96235.
8.61	96374.
8.59	96143.
8.16	86319.
8.20	86340.
8.18	85913.
7.83	78211.
7.84	78087.
7.86	78205.
7.47	69778.
7.48	69595.
7.50	69466.
7.22	62902.
7.29	63125.
7.29	62907.
6.82	51729.
6.80	50760.
6.84	49142.
6.38	41090.
6.35	40671.
6.36	40642.
5.89	33909.
5.88	33654.
5.89	33758.
5.40	27044.
5.36	26513.
5.34	26154.
4.67	18716.
4.64	18454.
4.63	18377.
3.90	12039.
3.85	11697.
3.81	11398.
2.98	6248.
2.75	5782.
2.68	5573.
1.79	2782.
1.63	2769.
11.12	122435.
11.08	122355.
11.04	122220.
10.62	112951.
10.43	110311.
10.44	110007.
10.15	103899.
10.24	107825.
10.25	106604.
9.68	96072.
9.69	96173.

ΔT_s (K)	q'' (W/m²)
9.73	97980.
9.26	87854.
9.27	88043.
9.31	89341.
8.86	80000.
8.87	80245.
8.90	80465.
8.52	72195.
8.55	72251.
8.59	72708.
8.09	62767.
8.11	60591.
8.12	60201.
7.69	52011.
7.66	51128.
7.62	50434.
7.19	43051.
7.20	42911.
7.23	42902.
6.67	35072.
6.58	33796.
6.56	33727.
6.01	27173.
6.05	27391.
6.11	28057.
5.43	20724.
5.38	20429.
5.39	20481.
4.68	14030.
4.57	13600.
4.56	13527.
3.68	8134.
3.50	7608.
3.48	7475.
1.75	2530.
1.65	2535.
11.15	112669.
11.11	112677.
11.07	112800.
10.74	106032.
10.68	105356.
10.66	105174.
10.31	97738.
10.25	96981.
10.22	96406.
9.82	88237.
9.79	87991.
9.82	88471.
9.37	79479.
9.38	79447.
9.36	79151.
8.96	71426.
8.99	71662.
9.02	72064.
8.58	63308.
8.60	63325.
8.64	63657.
8.12	54547.
8.13	54098.
8.16	54232.
7.62	45195.
7.63	45319.
7.62	44894.
7.05	36866.
7.06	36753.

ΔT_s (K)	q'' (W/m²)
7.06	36660.
6.64	31420.
6.59	31030.
6.60	30876.
5.77	22205.
5.76	22053.
5.73	21749.
5.13	16436.
5.06	16219.
5.08	16257.
4.15	9906.
4.02	9522.
4.05	9727.
2.86	4726.
2.68	4690.
2.63	4498.
1.65	2445.
1.56	2352.
10.73	110779.
10.66	110100.
10.62	109790.
10.31	103069.
10.32	103639.
10.34	104736.
9.83	94338.
9.85	95036.
9.88	95658.
9.39	86180.
9.40	86366.
9.41	87096.
8.94	77279.
8.96	77560.
8.98	77796.
8.58	69371.
8.58	69217.
8.61	69559.
8.10	59705.
8.09	59194.
8.11	59145.
7.68	51440.
7.73	51485.
7.71	51025.
7.22	42740.
7.21	42262.
7.18	41673.
6.78	35661.
6.76	35429.
6.73	35074.
6.06	26871.
5.98	26293.
6.00	26106.
5.43	20565.
5.35	19778.
5.33	19563.
4.85	15255.
4.78	14898.
4.76	14934.
3.86	8809.
3.73	8516.
3.71	8426.
2.43	3944.
2.33	3797.
2.28	3646.
1.62	2374.
1.54	2369.

ΔT_s (K)	q'' (W/m²)
10.98	113845.
10.94	113809.
10.86	112830.
10.50	105550.
10.49	105778.
10.44	105057.
10.10	97695.
10.12	98492.
10.07	97150.
9.66	88738.
9.66	89176.
9.69	89980.
9.26	80506.
9.25	80792.
9.33	82337.
8.88	73063.
8.90	73475.
8.93	73869.
8.45	64741.
8.44	64571.
8.48	64905.
7.99	55932.
7.98	55438.
8.00	55152.
7.54	47200.
7.54	47132.
7.52	46447.
6.97	38284.
6.93	37454.
6.96	37695.
6.49	31390.
6.49	31159.
6.51	31505.
5.84	23671.
5.81	23503.
5.82	23579.
5.18	17573.
5.14	17197.
5.12	17096.
4.09	9957.
3.99	9629.
3.96	9428.
2.80	4932.
2.70	4711.
2.67	4634.
1.72	2437.
1.62	2446.
1.60	2450.

R134a/RL681Al0 (98/2)
File: Al202.dat

ΔT_s (K)	q'' (W/m²)
10.28	114218.
10.15	112853.
10.11	113735.
9.55	108154.
9.50	107987.
9.52	108510.
9.08	98188.
9.12	98495.
9.17	98830.
8.75	89599.
8.76	89963.

8.80	90311.	6.22	37246.	8.72	113807.
8.39	81914.	5.87	30372.	8.74	114207.
8.38	82016.	5.79	30080.	8.82	115289.
8.41	82354.	5.76	30243.	8.51	106935.
8.01	74335.	5.30	21958.	8.59	107879.
8.03	74359.	5.26	21666.	8.63	108484.
8.04	74827.	5.24	21575.	8.20	96972.
7.62	65604.	4.81	15470.	8.21	97000.
7.60	65443.	4.77	15081.	8.28	97828.
7.60	65326.	4.75	14685.	7.96	89409.
7.17	56310.	4.13	9443.	7.92	88625.
7.15	55590.	4.12	9371.	7.92	89315.
7.15	55145.	4.09	9272.	7.55	80354.
6.81	48165.	2.95	3901.	7.58	80733.
6.77	47294.	2.79	3755.	7.59	80911.
6.78	46971.	9.26	128999.	7.19	70805.
6.32	39644.	9.28	131015.	7.22	71026.
6.28	39433.	9.29	132781.	7.22	70907.
6.29	39067.	8.96	126296.	6.92	63023.
5.83	31594.	8.95	127286.	6.90	62432.
5.83	31275.	8.93	128369.	6.91	62434.
5.80	30696.	8.46	116418.	6.47	51486.
5.35	22826.	8.44	115612.	6.50	51962.
5.31	22643.	8.39	114720.	6.52	51823.
5.29	22565.	8.18	109015.	6.23	43952.
4.72	14658.	8.15	108975.	6.25	44230.
4.67	14301.	8.14	108788.	6.30	44765.
4.61	13625.	7.86	99715.	5.85	34339.
4.11	9222.	7.85	99344.	5.77	33851.
4.00	8815.	7.73	95936.	5.73	33474.
3.93	8558.	7.57	91016.	5.44	27010.
2.78	3456.	7.57	91846.	5.40	26544.
2.64	3534.	7.62	92961.	5.41	26532.
10.88	116199.	7.20	81653.	4.96	18851.
10.79	120613.	7.25	82842.	4.94	18727.
10.74	121628.	7.28	83606.	4.94	18657.
10.22	112686.	6.91	73057.	4.30	10912.
10.18	113049.	6.87	72348.	4.22	10373.
10.18	113637.	6.90	72296.	4.15	9933.
9.75	105841.	6.58	64324.	3.29	5117.
9.73	106001.	6.59	64110.	3.17	4951.
9.72	106181.	6.60	64207.	10.42	116953.
9.22	97058.	6.19	53653.	10.43	117080.
9.19	97102.	6.19	52797.	10.40	117375.
9.19	97663.	6.10	50305.	9.98	109852.
8.79	89674.	5.88	45288.	9.90	109022.
8.77	89567.	5.91	45085.	9.91	108913.
8.76	89535.	5.93	45205.	9.46	101216.
8.34	80846.	5.54	35943.	9.46	100776.
8.32	80472.	5.54	36727.	9.45	100704.
8.30	80423.	5.51	37315.	9.09	93332.
7.90	71686.	5.13	28635.	9.07	92903.
7.87	70953.	5.09	28748.	9.07	92936.
7.82	69884.	5.05	28685.	8.66	84604.
7.49	62992.	4.71	20683.	8.66	84367.
7.48	62697.	4.66	20194.	8.67	84367.
7.50	63069.	4.65	19974.	8.24	75401.
7.15	55622.	4.10	12041.	8.22	75026.
7.15	55262.	4.02	11528.	7.79	66027.
7.20	55626.	4.02	11408.	7.75	65341.
6.78	47185.	3.16	5698.	7.81	65484.
6.76	46507.	3.11	5531.	7.59	58621.
6.76	46249.	9.09	123097.	7.60	58026.
6.29	37447.	9.03	122015.	7.66	58087.
6.26	37439.	8.90	119253.	7.26	50166.

7.27	50196.	7.43	58810.	7.43	59539.
7.29	50372.	7.06	50301.	7.43	59218.
6.77	41410.	7.07	50132.	7.48	59095.
6.73	40949.	7.08	49499.	7.00	49402.
6.74	40875.	6.61	41399.	6.97	48706.
6.24	34627.	6.55	41619.	6.97	48603.
6.24	34763.	6.55	41608.	6.55	41095.
6.25	34698.	6.15	33733.	6.44	40372.
5.57	26991.	6.09	33694.	6.40	39750.
5.53	27040.	6.07	33724.	6.09	33694.
5.57	27458.	5.65	25778.	6.06	33722.
4.91	19637.	5.58	25098.	6.05	33902.
4.87	19415.	5.57	24874.	5.63	26840.
4.85	19349.	5.15	18009.	5.58	26422.
4.28	12742.	5.08	17573.	5.58	26605.
4.21	12420.	5.05	17384.	5.12	18828.
4.19	12356.	4.40	10458.	5.07	18361.
3.20	5442.	4.32	10045.	5.07	18425.
3.12	5276.	4.27	9801.	4.24	9768.
10.71	125549.	3.41	4948.	4.21	9518.
10.55	126467.	3.29	4802.	4.19	9485.
10.49	126929.	10.68	127176.	3.25	4468.
10.01	119141.	10.49	125440.	3.08	4354.
9.94	118860.	10.10	120469.	3.03	4247.
9.96	119039.	9.98	119371.	1.85	2913.
9.62	111968.	9.98	120028.	1.69	3010.
9.63	111431.	10.00	120934.	1.66	2886.
9.64	111897.	9.66	111448.	2.99	3600.
9.22	102778.	9.72	112261.	2.78	3594.
9.21	102353.	9.76	113110.	2.72	3516.
9.20	102467.	9.24	102737.	2.04	2405.
8.86	95094.	9.27	102926.	1.95	2424.
8.83	94762.	9.29	102883.	1.90	2373.
8.83	94577.	8.85	93678.	2.97	3689.
8.45	85922.	8.81	93281.	2.87	3557.
8.45	85537.	8.80	92841.	2.18	2360.
8.41	84735.	8.43	85004.	2.06	2372.
8.03	75997.	8.42	84646.	2.86	3360.
8.00	75401.	8.43	85092.	2.61	3346.
8.02	75845.	8.08	77322.	2.50	3153.
7.73	68457.	8.09	77729.	2.39	2530.
7.76	69035.	8.12	77835.	2.22	2611.
7.81	69605.	7.73	68465.	2.19	2642.
7.39	59110.	7.75	68154.		
7.43	59116.	7.77	67640.		

Table 3 Number of test days and data points

Fluid (% mass fraction)	Number of days	Number of data points
Pure R134a $2.7 \text{ K} \leq \Delta T_s \leq 8.3 \text{ K}$	5	213
R134a/RL68H (99.5/0.5) $3.8 \text{ K} \leq \Delta T_s \leq 9.1 \text{ K}$	5	203
R134a/RL68H (99/1) $3.7 \text{ K} \leq \Delta T_s \leq 8.8 \text{ K}$	7	552
R134a/RL68H (98/2) $5.3 \text{ K} \leq \Delta T_s \leq 10.0 \text{ K}$	7	231
R134a/RL681AlO (99.5/0.5) $1.3 \text{ K} \leq \Delta T_s \leq 10.1 \text{ K}$	8	360
R134a/RL681AlO (99/1) $1.6 \text{ K} \leq \Delta T_s \leq 11.2 \text{ K}$	7	327
R134a/RL681AlO (98/2) $1.7 \text{ K} \leq \Delta T_s \leq 10.1 \text{ K}$	7	339

Table 4 Estimated parameters for cubic boiling curve fits for plain copper surface

$$\Delta T_s = A_0 + A_1 q'' + A_2 q''^2 + A_3 q''^3$$

ΔT_s in kelvin and q'' in W/m^2

Fluid	A_o	A_1	A_2	A_3
Pure R134a				
2.7 K $\leq \Delta T_s \leq$ 7.0 K	2.00560	3.50161×10^{-4}	-8.79342×10^{-9}	7.80541×10^{-14}
7.0 K $\leq \Delta T_s \leq$ 8.3 K	5.67285	4.38058×10^{-5}	-2.48689×10^{-10}	5.21295×10^{-16}
R134a/RL68H (99.5/0.5)				
3.8 K $\leq \Delta T_s \leq$ 7.6 K	2.99054	3.13110×10^{-4}	-8.37866×10^{-9}	8.39947×10^{-14}
7.6 K $\leq \Delta T_s \leq$ 9.1 K	5.94731	5.45098×10^{-5}	-4.07608×10^{-10}	1.38397×10^{-15}
R134a/RL68H (99/1)				
3.7 K $\leq \Delta T_s \leq$ 7.3 K	2.91330	2.58294×10^{-4}	-5.76127×10^{-9}	4.79952×10^{-14}
7.3 K $\leq \Delta T_s \leq$ 8.8 K	7.71778	-3.23665×10^{-5}	6.61861×10^{-10}	-2.67140×10^{-15}
R134a/RL68H (98/2)				
5.3 K $\leq \Delta T_s \leq$ 8.2 K	3.76719	2.51828×10^{-4}	-5.48232×10^{-9}	4.31318×10^{-14}
8.2 K $\leq \Delta T_s \leq$ 10.0 K	6.84177	1.55996×10^{-5}	2.78987×10^{-10}	-1.58215×10^{-15}
R134a/RL681AlO (99.5/0.5)				
1.3 K $\leq \Delta T_s \leq$ 6.2 K	0.602073	3.98702×10^{-4}	-1.18786×10^{-8}	1.42247×10^{-13}
6.2 K $\leq \Delta T_s \leq$ 10.1 K	3.52797	9.50689×10^{-5}	-5.28955×10^{-10}	1.44536×10^{-15}
R134a/RL681AlO (99/1)				
1.6 K $\leq \Delta T_s \leq$ 11.2 K	1.82058	1.99793×10^{-4}	-2.19331×10^{-9}	1.02134×10^{-14}
R134a/RL681AlO (99/2)				
1.7 K $\leq \Delta T_s \leq$ 4.0 K	0.143420	9.52658×10^{-4}	-7.62885×10^{-8}	2.04666×10^{-12}
4.0 K $\leq \Delta T_s \leq$ 10.1 K	3.21551	1.06794×10^{-4}	-9.03512×10^{-10}	4.03110×10^{-15}

Table 5 Residual standard deviation of ΔT_s

Fluid	(K)
Pure R134a	
$2.7\ \mathrm{K} \leq \Delta T_s \leq 7.0\ \mathrm{K}$	0.14
$7.0\ \mathrm{K} \leq \Delta T_s \leq 8.3\ \mathrm{K}$	0.06
R134a/RL68H (99.5/0.5)	
$3.8\ \mathrm{K} \leq \Delta T_s \leq 7.6\ \mathrm{K}$	0.09
$7.6\ \mathrm{K} \leq \Delta T_s \leq 9.1\ \mathrm{K}$	0.04
R134a/RL68H (99/1)	
$3.7\ \mathrm{K} \leq \Delta T_s \leq 7.3\ \mathrm{K}$	0.08
$7.3\ \mathrm{K} \leq \Delta T_s \leq 8.8\ \mathrm{K}$	0.09
R134a/RL68H (98/2)	
$5.3\ \mathrm{K} \leq \Delta T_s \leq 8.2\ \mathrm{K}$	0.09
$8.2\ \mathrm{K} \leq \Delta T_s \leq 10.0\ \mathrm{K}$	0.16
R134a/RL681AlO (99.5/0.5)	
$1.3\ \mathrm{K} \leq \Delta T_s \leq 6.2\ \mathrm{K}$	0.22
$6.2\ \mathrm{K} \leq \Delta T_s \leq 10.1\ \mathrm{K}$	0.25
R134a/RL681AlO (99/1)	
$1.6\ \mathrm{K} \leq \Delta T_s \leq 11.2\ \mathrm{K}$	0.49
R134a/RL681AlO (99/2)	
$1.7\ \mathrm{K} \leq \Delta T_s \leq 4.0\ \mathrm{K}$	0.22
$4.0\ \mathrm{K} \leq \Delta T_s \leq 10.1\ \mathrm{K}$	0.44

Table 6 Average magnitude of 95 % multi-use confidence interval for mean T_w -T_s (K)

Fluid	U (K)
Pure R134a	
2.7 K ≤ ΔT_s ≤ 7.0 K	0.10
7.0 K ≤ ΔT_s ≤ 8.3 K	0.03
R134a/RL68H (99.5/0.5)	
3.8 K ≤ ΔT_s ≤ 7.6 K	0.07
7.6 K ≤ ΔT_s ≤ 9.1 K	0.03
R134a/RL68H (99/1)	
3.7 K ≤ ΔT_s ≤ 7.3 K	0.04
7.3 K ≤ ΔT_s ≤ 8.8 K	0.03
R134a/RL68H (98/2)	
5.3 K ≤ ΔT_s ≤ 8.2 K	0.08
8.2 K ≤ ΔT_s ≤ 10.0 K	0.08
R134a/RL681AlO (99.5/0.5)	
1.3 K ≤ ΔT_s ≤ 6.2 K	0.14
6.2 K ≤ ΔT_s ≤ 10.1 K	0.11
R134a/RL681AlO (99/1)	
1.6 K ≤ ΔT_s ≤ 11.2 K	0.18
R134a/RL681AlO (99/2)	
1.7 K ≤ ΔT_s ≤ 4.0 K	0.22
4.0 K ≤ ΔT_s ≤ 10.1 K	0.16

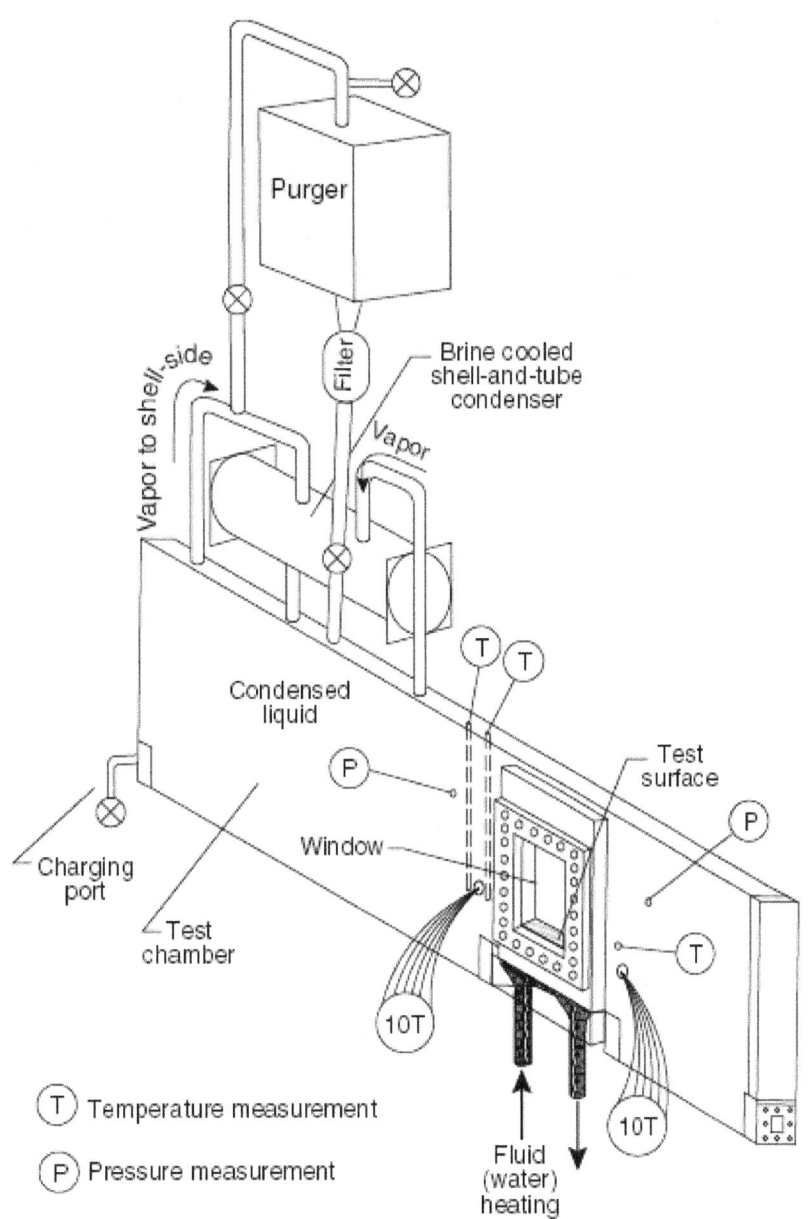

Fig. 1 Schematic of test apparatus

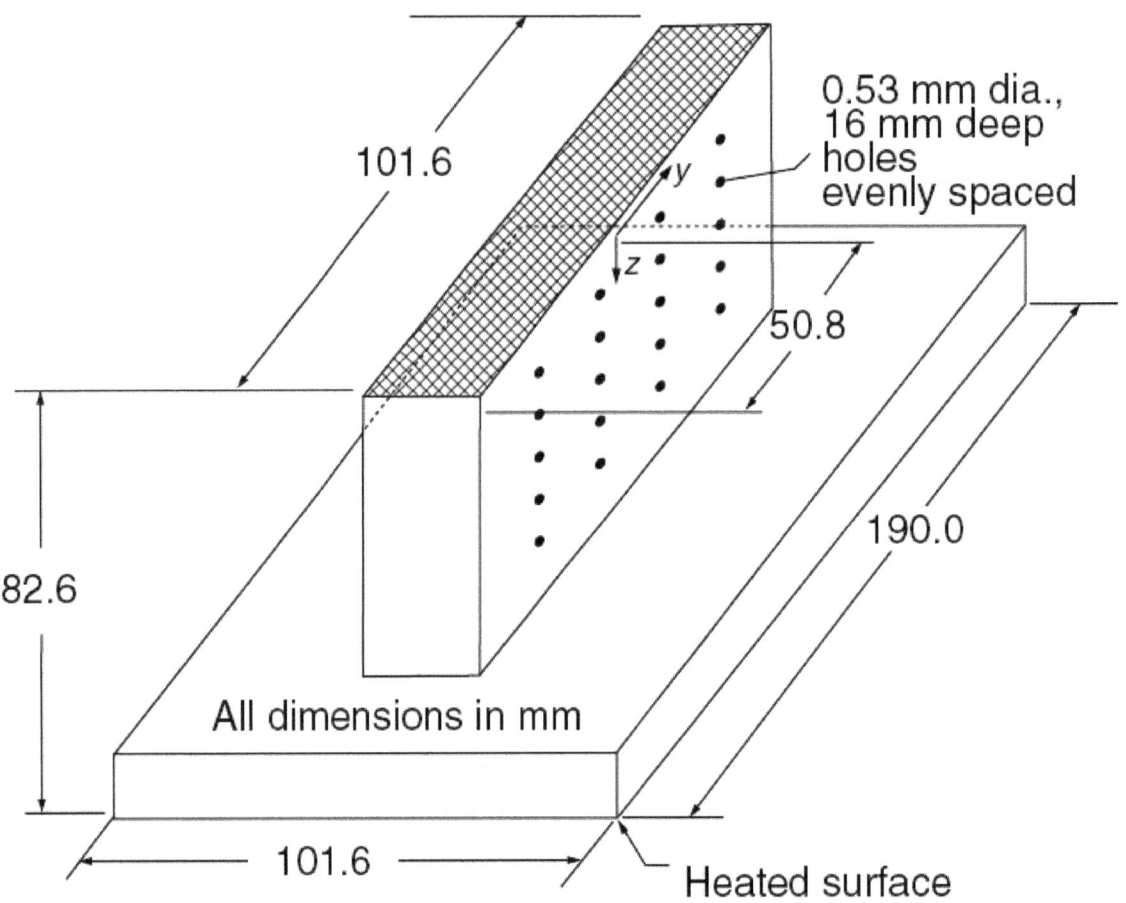

101.6

0.53 mm dia.,
16 mm deep
holes
evenly spaced

y

z

50.8

190.0

82.6

All dimensions in mm

101.6

Heated surface

**Fig. 2 OFHC copper flat test plate with cross-hatched surface and thermocouple
coordinate system**

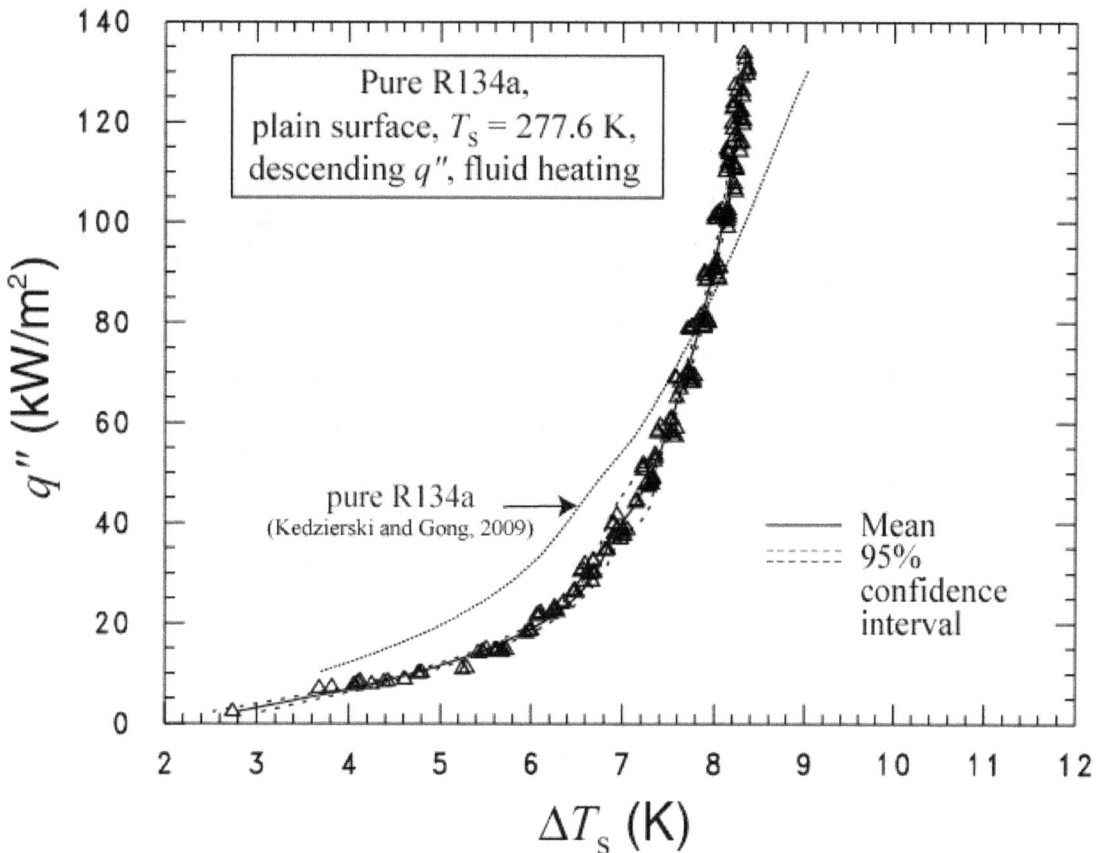

Fig. 3 Pure R134a boiling curve for plain surface

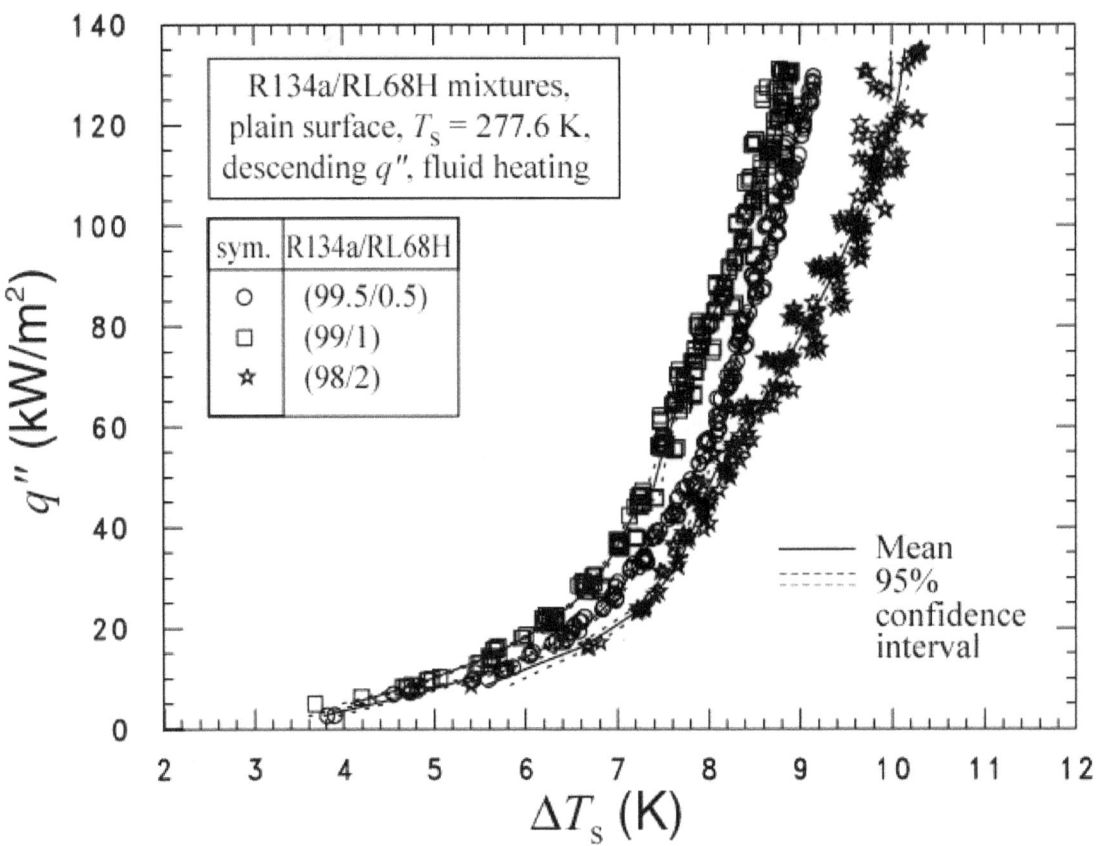

Fig. 4 R134a/RL68H mixtures boiling curves

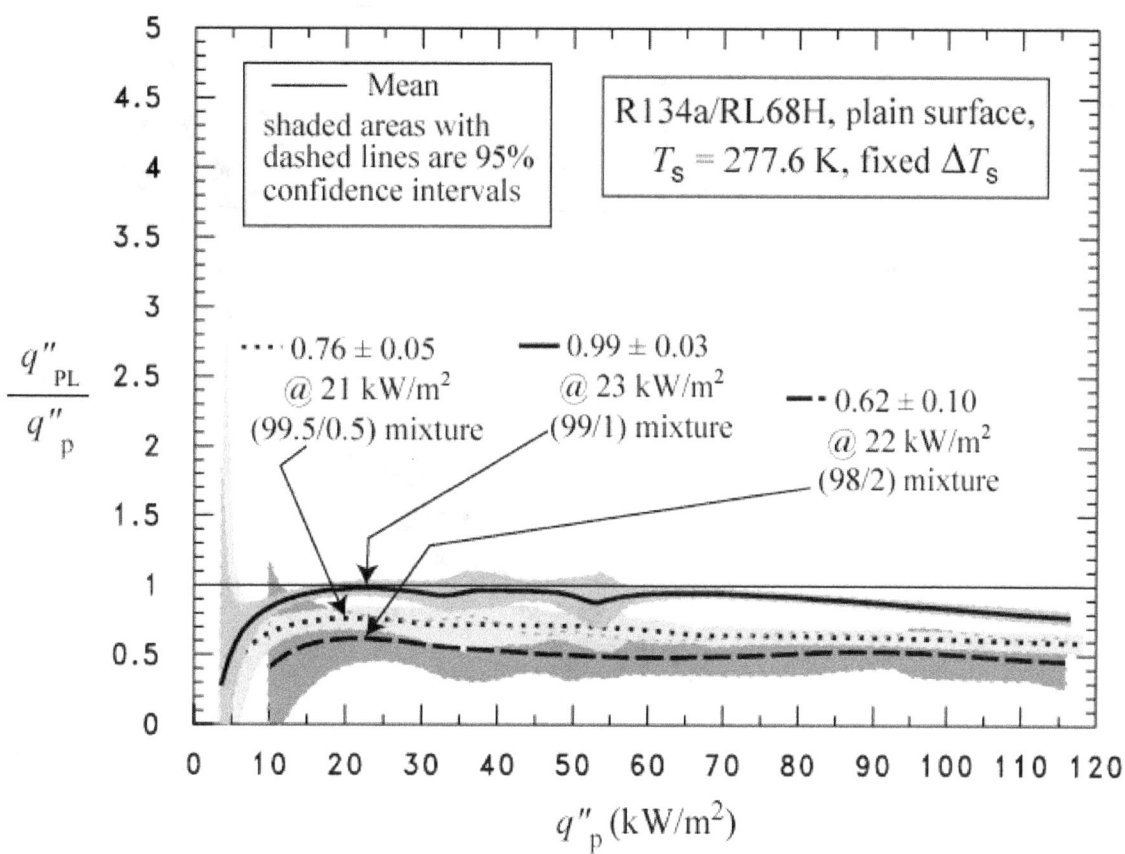

Fig. 5 Boiling heat flux of R134a/RL68H mixture relative to that of pure R134a

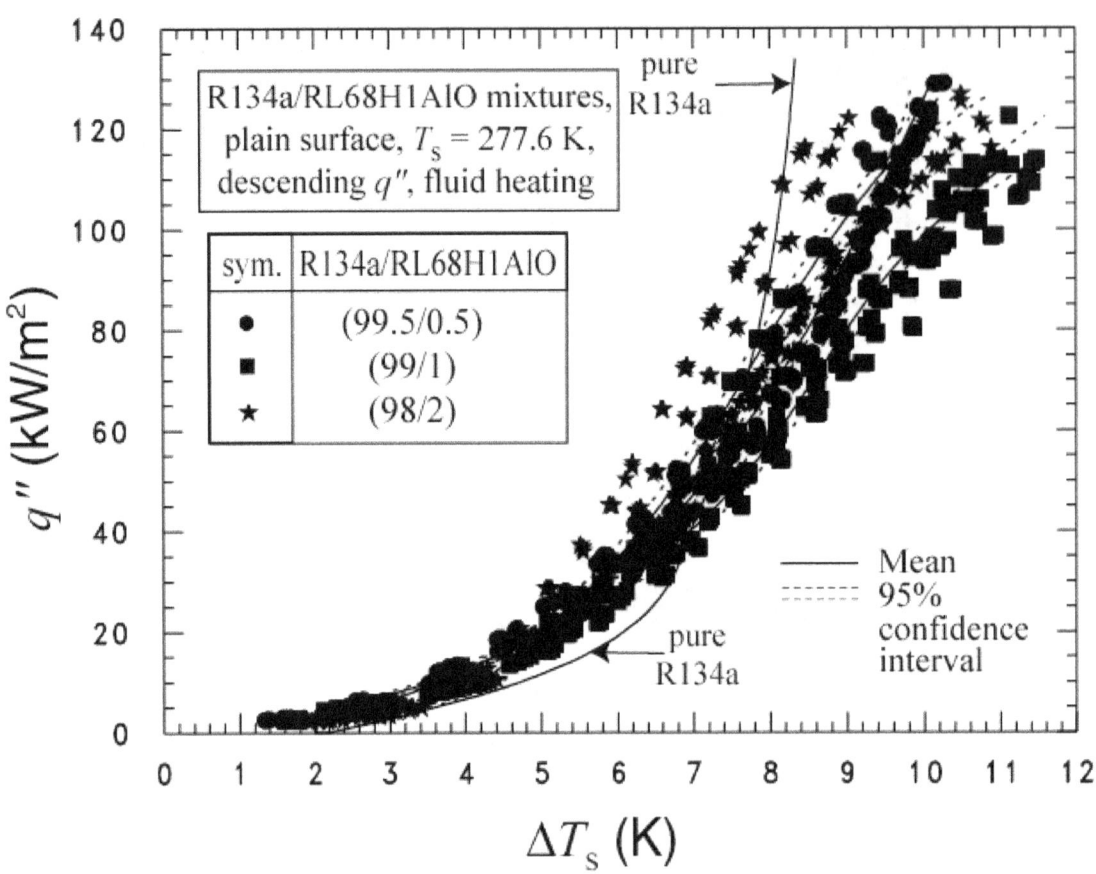

Fig. 6 R134a/RL681AlO mixtures boiling curves

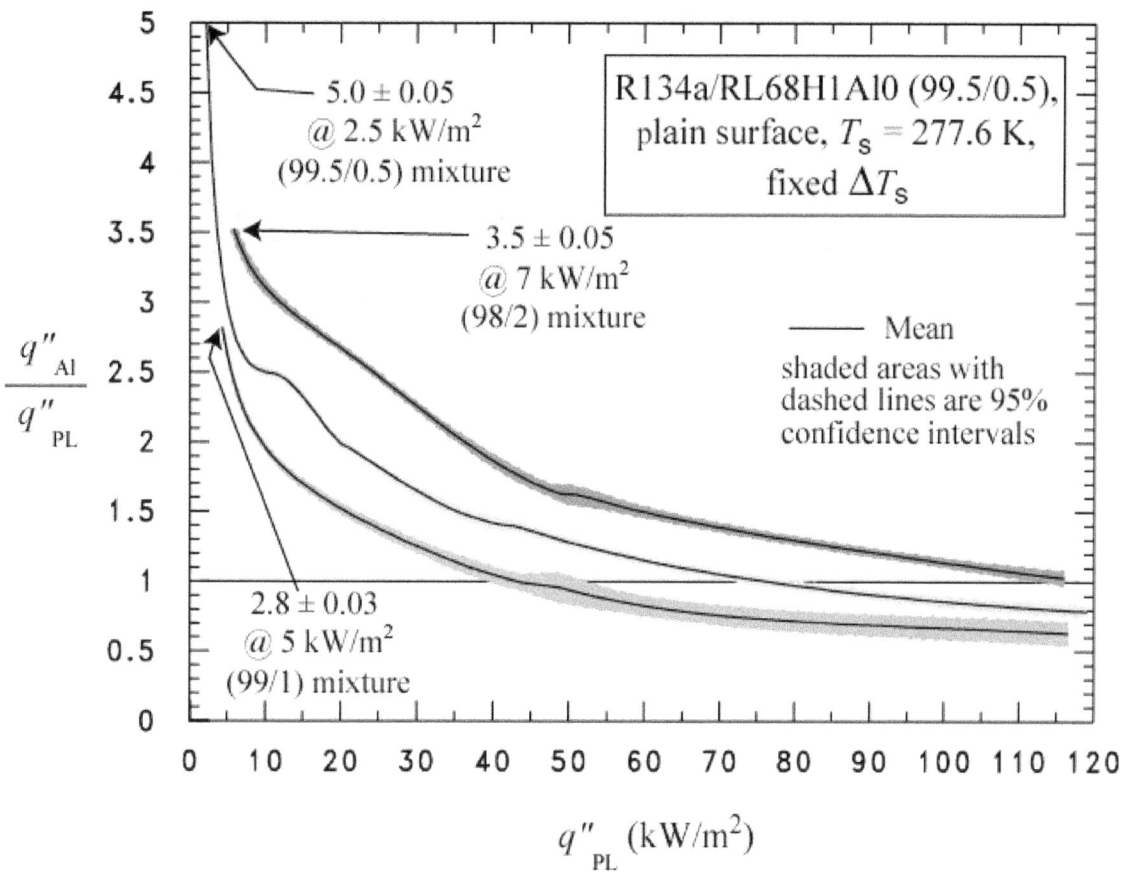

Fig. 7 Boiling heat flux of R134a/RL68H1AlO mixtures relative to that of R134a/RL68H without nanoparticles

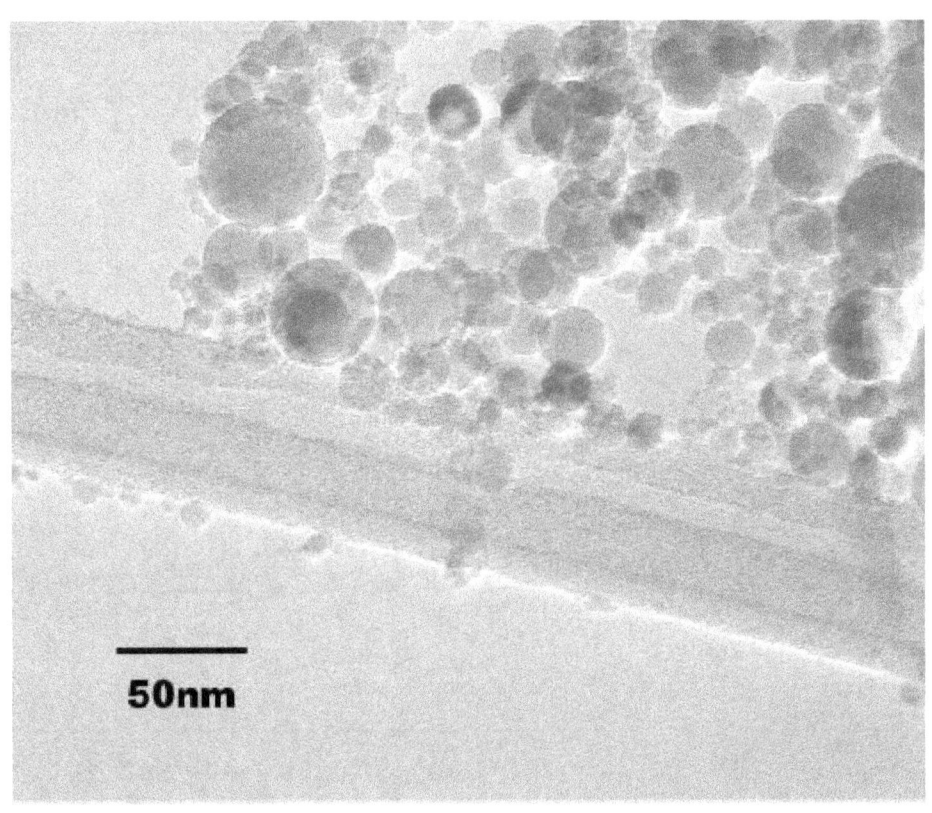

Fig. 8 TEM of Al₂O₃ nanolubricant (Sarkas, 2009)

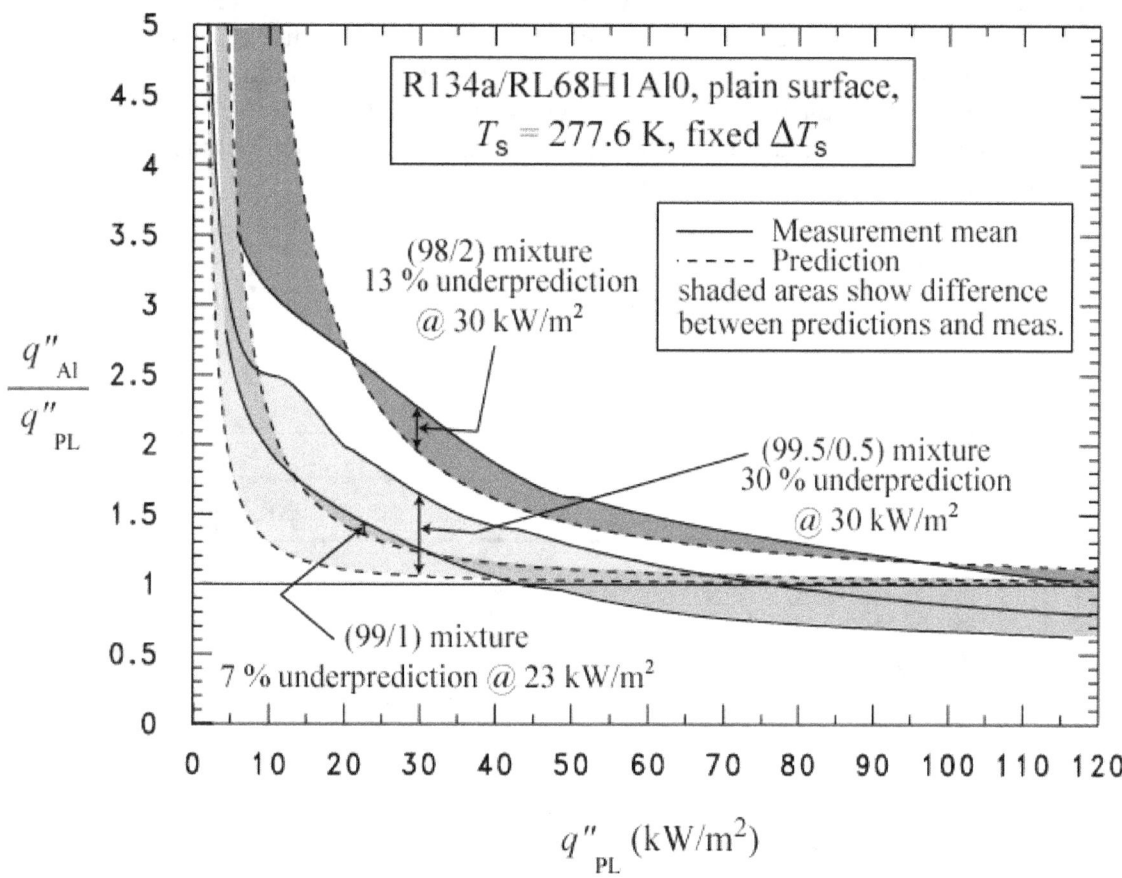

Fig. 9 Comparison of measured to predicted boiling heat flux ratios for the R134a/RL68H1AlO mixtures

APPENDIX A: UNCERTAINTIES

Figure A.1 shows the relative (percent) uncertainty of the heat flux ($U_{q''}$) as a function of the heat flux. Figure A.2 shows the uncertainty of the wall temperature as a function of heat flux. The uncertainties shown in Figs. A.1 and A.2 are "within-run uncertainties." These do not include the uncertainties due to "between-run effects" or differences observed between tests taken on different days. The "within-run uncertainties" include only the random effects and uncertainties associated with one particular test. All other uncertainties reported in this study are "between-run uncertainties" which include all random effects such as surface past history or seeding.

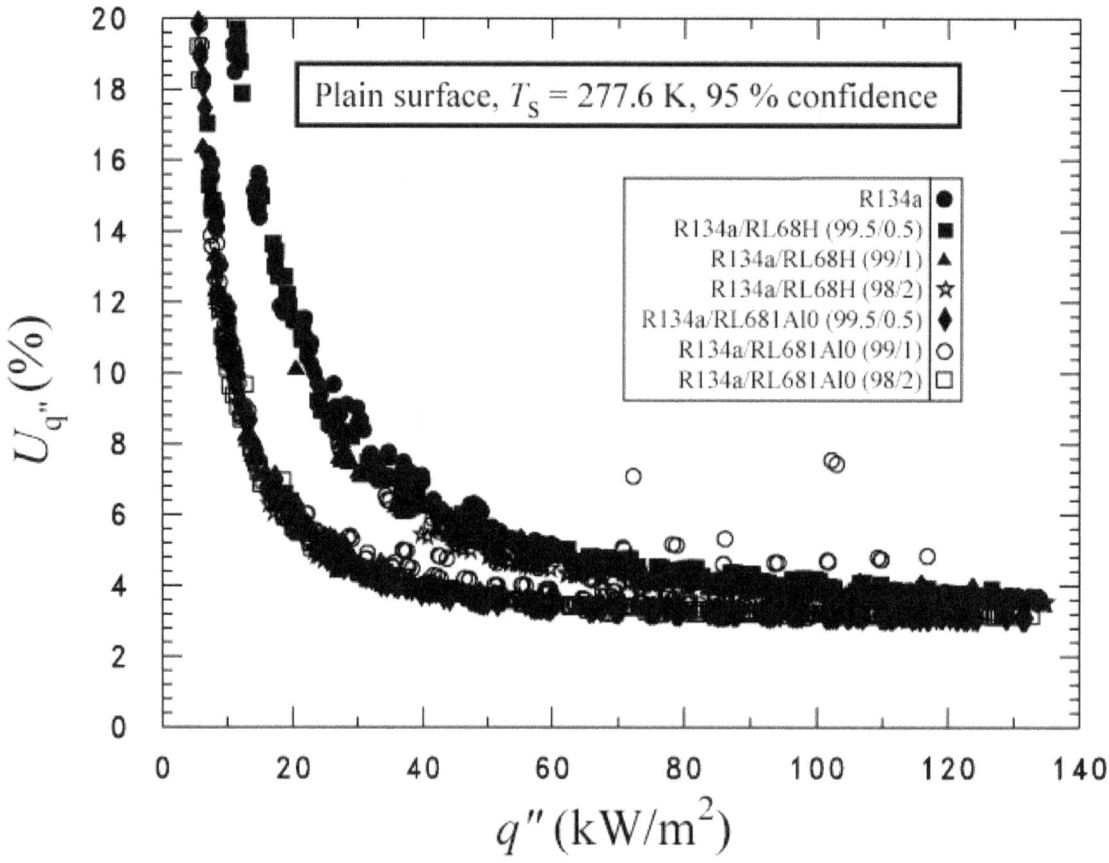

Fig. A.1 Expanded relative uncertainty in the heat flux of the surface at the 95 % confidence level

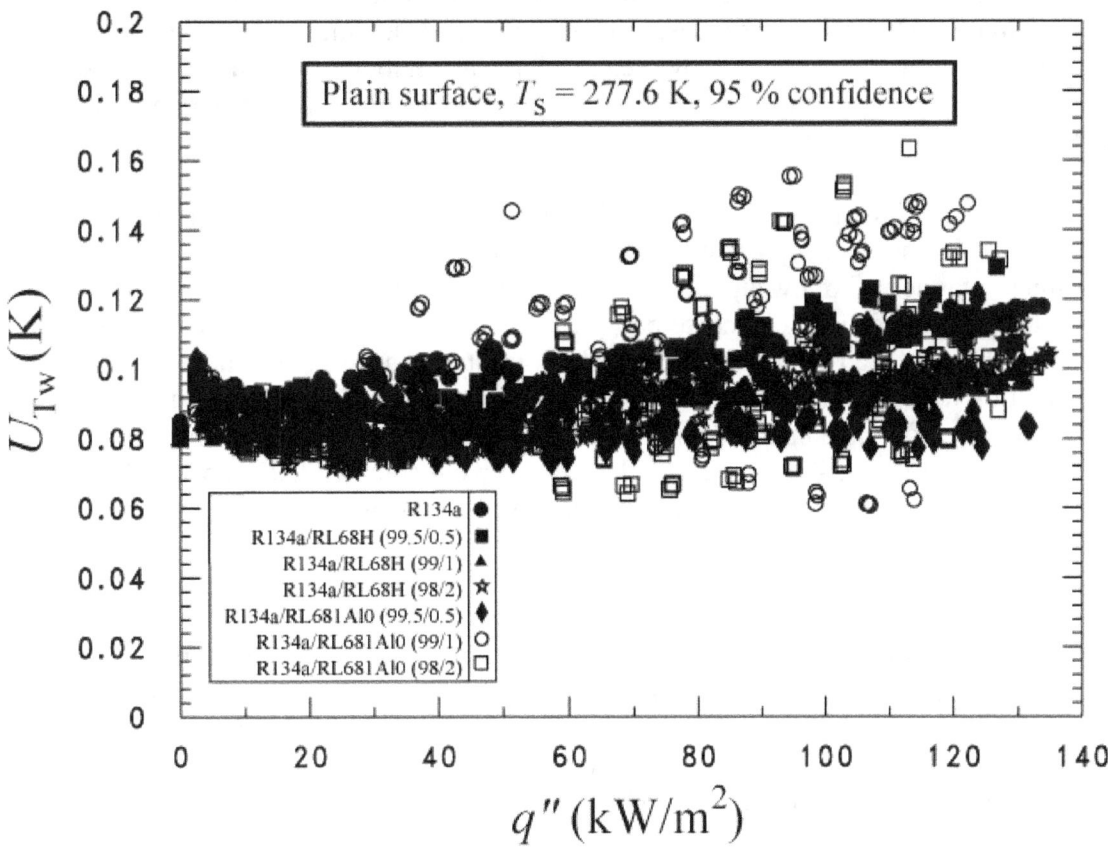

**Fig. A.2 Expanded uncertainty in the temperature of the surface at the 95 %
confidence level**

APPENDIX B: LIQUID VISCOSITY AND DENSITY MEASUREMENTS

This appendix presents the liquid density and viscosity measurements for the 5.6 % mass fraction (1.6 % volume fraction) Al_2O_3 nanolubricant. A Stabinger Viscometer was used to measure the dynamic viscosity and the density of the liquid nanolubricant at approximately 313.15 K. The measurements were made at atmospheric pressure at an approximate altitude of 137 m above sea level at Gaithersburg, Maryland, USA. The viscometer uses a vibrating U-tube to determine the density. The kinematic viscosity reported here is obtained by dividing the dynamic viscosity by the density.

The operation principle of the Stabinger Viscometer relies on rotating concentric cylinders. The liquid sample is contained in the annulus of a concentric cylinder where the inner cylinder is hollow and of less mass than the sample. This allows the inner cylinder to float freely and be centered by centrifugal forces in the sample when the outer cylinder is spun by a rotating magnetic field. Viscous shear forces on the liquid transfer the rotation to the inner cylinder. Measurements on the inner cylinder are used to calculate the difference in speed and torque between the outer and inner cylinder, and thus, the dynamic viscosity. Wasp et al. (1977) have recommended concentric cylinder viscometers for use with solid-liquid suspensions. All calculations are internal to the instrument.

The viscometer-manufacturer quoted uncertainty for the 95 % confidence level for the kinematic viscosity and the density was $\pm$ 0.35 % and $\pm$ 0.5 kg$\cdot$m^{-3}, respectively. The viscometer was used to measure the density and viscosity of a calibration fluid with a nominal viscosity and density at 293.15 K of 1320 mm$^2\cdot$s^{-1} and 845.4 kg$\cdot$m^{-3}, respectively. Residuals between the measurements and the calibration standard over the same temperature range of this study were within the quoted specifications of the manufacturer.

Table B.1 presents the liquid density and viscosity measurements for the 5.6 % mass fraction Al_2O_3 nanolubricant (RL68H1AlO). The average liquid viscosity of RL68H1AlO at 313.15 K is 73 mm$^2\cdot$s^{-1} $\pm$ 1 mm$^2\cdot$s^{-1}. The average liquid density of RL68H1AlO at 313.15 K is 1006.5 kg$\cdot$m^{-3} $\pm$ 0.5 kg$\cdot$m^{-3}.

Comparison of the nanolubricant density measurements to the recommended mixture equation for suspensions (Wasp et al., 1977):

$$\frac{1}{\rho_{nL}} = \frac{x_{nL}}{\rho_{np}} + \frac{1-x_{nL}}{\rho_L} \, , \tag{B.1}$$

results in a predited of the nanolubricant mixture density (ρ_{nL}) of 1005.2 kg$\cdot$m^{-3}, which is within 0.1 % of the measured value. This occurred when a Al_2O_3 density[4] of 3600 kg$\cdot$m^{-3} was used for the density of the solid nanoparticles (ρ_{np}) along with the correlated density

---

[4] Bill Turner, Nanophase Technologies, Private communications, March 3, 2009.

value for the pure lubricant (ρ_L = 963.98 kg·m^{-3}) at 313.15 K from Kedzierski (2009c). The mass fraction of the nanolubricant is x_{nL}.

Comparison of the viscosity measurements to the Stokes-Einstein equation (Einstein, 1956) for a dilute (ϕ < 0.6) nanofluid with spherical nanoparticles,

$$\frac{v_{nL}\rho_{nL}}{v_L\rho_L} = 1 + (5/2)\phi \tag{B.2}$$

where ϕ is the volume fraction of the nanoparticles in the nanolubricant. The viscosity of the pure RL68H lubricant (v_L) at 313.15 K was taken from Kedzierski (2009c) as approximately 64.45 mm^2·s^{-1}. Equation B.2 produces a viscosity of approximately 64.19 mm^2·s^{-1}, which is 12 % smaller than the measured value.

The measured nanolubricant viscosity was also compared to the correlation of the liquid kinematic viscosity that Kedzierski (2009c) developed for RL68H and CuO nanolubricants as a function of the liquid density and the normalized temperature ($T_r = T/273.15$ K):

$$v_{nL}[\text{mm}^2 \bullet \text{s}^{-1}] = 2.02 \times 10^{-5} \left(\frac{\rho_{nL}}{1000\,[\text{kg} \cdot \text{m}^3]} \right)^{3.8} \exp\left(\frac{17.2}{T_r} \right) \tag{B.3}$$

Here the kinematic viscosity has units of mm^2·s^{-1} while the density (ρ_{nL}) has units of kg·m^{-3}. Equation B.3 gives a visosity of 67.88 mm^2·s^{-1}, which is 7 % smaller than the measured value. The good agreement is surprising considering the size and material difference between the CuO nanoparticles (35 nm) for which the correlation was developed and the aluminum oxided (10 nm) nanoparticles being predicted.

Table B.1 RL68H1AlO liquid viscosity and density measurements

T (°C)	v_{nL} (mm^2/s)	ρ_{nL} (kg/m^3)
40	74.847	1006.1
40	74.881	1006.0
40	74.889	1006.0
40	74.884	1006.0
40	74.872	1006.1
40	74.865	1006.1
40	74.857	1006.1
40	74.850	1006.1
40	74.837	1006.1
40	74.831	1006.1
40	74.826	1006.1
40	74.811	1006.1
40	70.874	1006.9
40	70.763	1006.9
40	70.702	1006.9
40	70.615	1006.9
40	70.565	1007.0
40	70.474	1007.0
40	70.406	1007.0
40	70.316	1007.1
40	70.223	1007.1
40	70.137	1007.1
40	70.039	1007.2
40	69.947	1007.2

APPENDIX C: LIQUID THERMAL CONDUCTIVITY MEASUREMENTS

This appendix presents the liquid thermal conductivity measurements for the 5.6 % mass fraction (1.6 % volume fraction) Al_2O_3 nanolubricant. A transient line-source technique (Roder et al., 2000) was used to measure the thermal conductivity of the liquid nanolubricant at an average temperature of 297.07 K ± 0.13 K. According to Vadasz (2006) the transient line-source technique is the most accurate for measuring the thermal conductivity of nanofluids as long as test duration is short enough to minimize natural convection, but long enough to minimize dual-phase-lagging effects. Lee et al. (1999) estimate that a measurement that is less than 5 s in duration is less likely to induce natural convection from the wire. The dual-phase-lagging effect arises from the difference in the heat capacities of the base-fluid and the solid particles, which results in a *"delayed response to any temperature variation in the neighboring fluid"* (Vadasz, 2006). Although the dual-phase-lagging effect was not calculated for the present measurements, example cases given by Vadasz (2006) shows that this effect is minimized for measurement durations of 5 s.

From the above, it appears that a measurement duration of approximately 5 s is preferred in order to minimize errors using the transient line-source technique. The present measurements were made with a packaged instrument (KD2 Pro) that has a pre-determined measurement duration of approximately 30 s. The average liquid thermal conductivity of the 5.6 % mass fraction Al_2O_3 nanolubricant from the measurements presented in Table C.1 was 0.138 W/m·K ± 0.0003 W/m·K at 23.92 °C ± 0.13 °C.

The thermal conductivity of a nanolubricant (k_{nL}) can be estimated from the volume fraction (ϕ), the thermal conductivity of the base-lubricant (k_L), and the thermal conductivity of the nanoparticles (k_{np}) as (Wasp, 1977):

$$k_{nL} = k_L \left(\frac{\dfrac{k_{np}}{k_L} + 2 - 2\phi\left(1 - \dfrac{k_{np}}{k_L}\right)}{\dfrac{k_{np}}{k_L} + 2 + \phi\left(1 - \dfrac{k_{np}}{k_L}\right)} \right) \tag{C.1}$$

The thermal conductivity of the base-lubricant is approximately 0.132 W/m·K (Kedzierski and Gong, 2009). The k_{np} was obtained from Sibley et al. (1958) for aluminum oxide (30 W/m·K). Using the above thermal conductivities and a volume fraction of 0.016, eq. (C.1) gives a value of 0.1383 W/m·K, which is within the uncertainty of the measurement. Considering the agreement between the prediction and the measurement, the potential measurement errors due to natural convection as induced by a long measurement time appear to be insignificant.

Table C.1 RL68H1AlO liquid thermal conductivity measurements

T (°C)	k_{nL} (W/m·K)
23.57	0.138
23.75	0.138
23.62	0.137
23.64	0.138
23.74	0.138
23.80	0.138
23.92	0.138
23.95	0.138
23.99	0.138
24.16	0.139
24.17	0.138
24.16	0.137
24.16	0.138
24.21	0.139

www.ingramcontent.com/pod-product-compliance
Lightning Source LLC
Chambersburg PA
CBHW081900170526

45167CB00007B/3086